Aba – The Glory and the Torment

The Life of Dr. Immanuel Velikovsky

Ruth Velikovsky Sharon, Ph.D.

New, revised edition (2010)
Second Printing

Original edition 1995

Copyright © 1995 by Ruth V. Sharon

All rights reserved. No part of this book may be reproduced or transmitted in any form or by any means, electronic or mechanical, including photocopying, recording, or by any information storage and retrieval system, without permission in writing from the copyright owner, except by reviewers who may quote brief passages to be printed in a magazine or newspaper.

Cover design based on a sculpture by Elisheva Velikovsky

Published by   Paradigma Ltd.
                          Internet: www.paradigma-publishing.com
                          e-mail: info@paradigma-publishing.com

ISBN 978-1-906833-20-6

Contents

Acknowledgements	5
Introduction	7
Aba's Early Life	15
Sabta and Saba (Aba's Parents)	23
Ima	35
Living in Palestine	53
Aba and Psychoanalysis	65
The United States	101
Ima's Sculpture	113
Aba – The Observer	121
The Battle	125
Depression	163
Better Times	169
Aba's Death	185
Ima at the Helm	191
Aba's Legacy	201
Immanuel Velikovsky's Interdisciplinary Synthesis	211
The Books	219
Velikovsky's Solar System	277
Psychoanalytic Lecture	283
Around the Subject	303

Acknowledgements

Naomi, Rafael and Carmel who loved and were loved by Aba and Ima, and who shared in their glory and their torment.

Dr. Marilyn Frasier's talent and knowledge made her contribution and research invaluable in completing this book.

Hanna Lavigne's devotion to this project was constant.

Dr. John Seed's dedication to healing body and mind, is pervasive.

Introduction

Aba and Ima on the Carmel

"I regard my meeting Elisheva as the greatest luck I had in my life. The nobleness of her character, her femininity, her honesty, her self-denials – all is before me through now thirty-six years as an unending blessing."

Those flowers were the sweetest thing that ever happened to me. That must be the only case in the world that *you* sent *me* flowers, where it should have been the other way around.

(Aba sent flowers to my mother on HIS birthday)

Ima and Aba in Princeton

Introduction

When I think of my father, a towering man with wavy white hair, frail and shaken at the end, I feel pity for the difficult times he experienced, and for the loss this world has suffered, of which it is still for the greatest part unaware.

No one wants to be told that all his work and research during a lifetime is wrong. My father did that to scientists in many fields, and they did it to him.

By eliminating Velikovsky and subjecting him to ridicule the scientific establishment in physics, geology, history, etc., could remain "intact."

Giving credence to Velikovsky – one man – would mean destroying the credence of volumes and volumes. It was a dangerous situation for my father to have gotten himself into. The magnitude of his genius and his brilliance met with an equal measure of disdain, for he presented an overwhelming threat. His creativity and insight were due to a most unusual mind: an educated, curious psychoanalytic mind, ambitious and persistent, and eternally optimistic, except during bouts with depression.

As time passed, many of my father's once so-called outrageous claims have been absorbed into the general body of knowledge and textbooks, without giving him credit, much less apologizing. *The Test of Time*, one of his last books (still unpublished), recounts what was believed in 1950, what *Worlds in Collision* stated in 1950, and what has been accepted since due in part to space probes.

To the end of his life Aba[1] said, proudly, that he would not change one word in *Worlds in Collision*. He was confident that his theories, research and conclusions were correct although never claiming to be infallible. Vindication mounted with space exploration – but he remained the pariah among scientists to the end. How could one man be right about so much – a real embarrassment.

My father was tormented by the scientific establishment. It took its toll, but only his close friends and family knew it. He had been analyzed by Wilhelm Stekel, an analysis which was of brief three months' duration. My father, a highly complicated man with a virtually unanalyzed psyche, ventured into a dangerous area, telling scientists they were wrong. Expecting some criticism and resistance, he was not even remotely prepared for the onslaught of the scientific establishment. Thinking that it would take ten years for his theories to be accepted, he did not expect to be trampled on without let-up until his death 29 years after the publication of *Worlds in Collision*.

My father never attacked his enemies, no matter what tactics they resorted to to undermine his works and character. Accused of doing "more damage than prostitution and communism combined", paid ads for his books were refused in scientific journals. Labeled a charlatan, the depressive abyss in between years of brilliance became more frequent and took a heavy toll on an aging body. A proud man of great integrity, married to a woman of impeccable honesty, to be accused of dishonesty was painful. Eternally optimistic except during bouts with depression, his hope kept him alive.

Although secretly hoping for the Nobel Prize, he outwardly accepted his lot of castigation.

---

[1] "Aba" means "Daddy" in Hebrew.

Aba
Photo by Noveck

Introduction

Whenever I hear the words "Star Wars", I wonder if any scientists recognize the implication of *Worlds in Collision* and my father's posthumously published *Mankind in Amnesia,* which explained in psychological terms the unwillingness and inability of man to address its past as recounted in *Worlds in Collision* (and in other titles). As an amnesia victim who seeks to relive the experience erased from conscious memory, so does mankind try to relive the traumatic past this planet suffered from extraterrestrial forces, causing upheavals on earth which destroyed much of mankind and disrupted evolution. So frightening were these events, that although there have existed written records of ancients from around the world, the direct language and clues have been disregarded until my father, almost like a detective, with great personal peril, put together *Worlds in Collision*. Understanding what men of many generations have overlooked, my father brought to consciousness what had been laying dormant.

Few theories met with such vituperative attempts at silencing of the theoretician as the Velikovsky theory. *Stargazers and Gravediggers* – the watergate of science, published posthumously in 1983, went unreviewed and was soon remaindered.

> To Dr. Velikovsky, disagreement regarding facts and theories was integral to the scientific enterprise; he expected his views to be met with dissent; constructing them as working hypothesis, he hoped that others in the field might help him to get them tested by observation and experiment. He did not expect that soi-disant scientists would, without reading and reflection, blacken his reputation and libel his character because of his scholarly views.
>
> Horace M. Kallen: »Shapley, Velikovsky and The Scientific Spirit«
> in *Velikovsky Reconsidered*

Should the Nobel prize committee one day decide to posthumously award its prize to my father, human endurance should be part of the award. My father corrected many men and women in many fields of science – a crime compounded by his ability to write well along with pursuing the opportunity to present his theories to the general public. Had his theories been published in a scientific periodical he would have been proper. But there was nothing proper about my father's theories nor his approach to their publication, nor my father's mind. Infinitely creative, able to cross hedges and make fun of all science for its nonsensical errors and its holier-than-thou attitude – scientists had trouble protecting their territory and attacked his theories without reading his books, and others quoted those who had not read. The reaction of jealous minds of mediocre ability is to slam doors in the face of the outrageously talented – thus permitting mediocrity to reign, retarding progress, and putting us all in jeopardy.

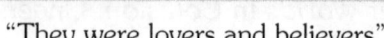
"They were lovers and believers"

"Immanuel Son of Simon Yehiel"

"Elisheva Daughter of Tuviah"

Is it all going to end suddenly for all of us? Are the stars going to be "at war" once more as described in *Worlds in Collision*? My father did not predict doom from extraterrestrial agents. He did predict, however, terrible danger, as described in *Mankind in Amnesia*, of reliving the trauma with a man-made nuclear holocaust. "Star Wars" has nothing to do with stars; it does, however, point to an unconscious frightening connection between earth's earlier experience with the stars and what politicians today call "Star Wars". Why the term "Star Wars"? It involves missiles in space, not the stars. "Those who do not remember the past are condemned to repeat it" – a quote originated by the Harvard philosopher George Santayana.

My father eventually realized that recognition and vindication would elude him in his lifetime. He often told me not to feel badly once he was vindicated after his lifetime. However, to have lived into the 90's and have seen that a silence and remaindering of his books has settled over his work would have further depressed him. Had he lived, however, perhaps there would not have been this silence. He is dead and my mother is dead, and there is no need on the part of the scientific establishment to flex its muscles any longer. The last nail in his coffin has been secured. However, his dedicated followers continue the work. My father did all he could to educate us and alert us. A man of vision – often tormented – he is at peace.

In Toms River, near their summer home on Pelican Island, there is a four plot site in the cemetery where there is one tombstone at the head of two graves, my father's and my mother's. My mother had made that selection for she was part of her husband in life, and in death, I believe, she wanted to remain part of his legacy. A two grave site space in front of their graves was to remain vacant, for I suspect my father knew that one day, far into the future, people would want to visit his gravesite.

Aba's Early Life

Daniel (Lelia), Alexander-Lev and Immanuel

Aba's Early Life

Immanuel Velikovsky was born on June 10th, 1895 in Russia in a cottage in the hills surrounding Vitebsk, birthplace of the artist Marc Chagall. About the same time period, Theodore Hertzel was writing his book *The Jewish State*, Freud had begun to write his *Interpretation of Dreams*, and in science, X-rays had been discovered. My father had two older brothers, Daniel and Alexander. In his unpublished autobiography, my father described vivid memories of his childhood in Vitebsk before the age of three where he reconstructed the town's map many years later.

Late in the year 1900 Immanuel's mother, father and Daniel left for Moscow to establish an import company and their new home. Immanuel and Alexander stayed in Vitebsk with their uncle, Israel, and a governess who tutored them in Russian and German. Soon after, in November of 1901 they joined their family in their new apartment in the most modern building in Moscow. My father's mother helped in the family business, but devoted most of her time ensuring that her sons would be educated. This included tutoring in French, which was required to enter the best school, the Medvednikov Gymnasium, in Moscow. They were also tutored in Russian, mathematics and in Hebrew by the only Hebrew teacher in Moscow. The government allowed Jews to comprise three percent of the student body, so when there was only one Jewish vacancy, his parents decided that Daniel, the eldest brother, should attend. My father was admitted the next year and excelled in Latin, Russian, history and mathematics and was editor of the class journal. His brother Alexander, who excelled in chemistry, later became a chemist and won the Lenin prize for his work.

During his youth my father traveled extensively. In the summer of 1912, one year before graduation, he made his first trip to Palestine. His father, never having been there, reminded him to kiss the earth for him. After visiting the Wailing Wall my father wrote a poem, *At the Wailing Wall*, which was published in *Razsvet*, a Zionist Organization weekly publication. The following year, 1913, he graduated with one of seven gold medals equivalent to "summa cum laude" from Medvednikov Gymnasium. After graduation he made another sojourn, this time to Finland with his two brothers.

Due to a new lottery system my father was denied admittance to Moscow University, but did not regret it and started making plans to study abroad. While medicine was not his choice of careers, his mother was insistent on it, and so my father attended Montpellier University in the south of France. Opthalmology interested him and he formulated a hypothesis that myopia could be reduced by cutting off part of the lens which would reduce refraction.

While at Montpellier he and a friend organized a student Zionist group. Unhappy with the strict academic environment, they left school and journeyed through Egypt to Palestine where they worked as laborers on a plantation in Rohoboth and later at Ruhama, which was founded by the group Sherith Israel, which his father had organized. My father's mother was upset by his decision, but his father was enthusiastic and supportive of his choice. Nevertheless, due to the boredom and the poverty of the conditions there Aba and his friend returned to Moscow.

Soon after, he entered the premedical program at the University of Edinburgh where he studied with visiting professor of philosophy, Henri Bergson. During his summer vacation at home in Moscow World War I began and my father was stranded in Russia. He spent the next two years studying at the Commercial Institute, a university that was started by the former rector and professors at Moscow University, who left after the government threatened to take away their autonomy. There my father studied law, economics, and his favorite subject, ancient history. Because of the lack of medical faculty, the education minister granted him permission to take medical courses as an intern student at Moscow University during the 1915-1916 school year. Working hard to accelerate his progress in school took toll on his health. My father contracted two types of pneumonia and had to spend several months recuperating in the Crimea. During that time he authored *The Third Exodus*, a pamphlet on religious Zionism that urged nations to create a Jewish state in Palestine. When he returned to Moscow in October 1917, to resume his medial studies, including a course in psychiatry, the Bolsheviks had taken power.

He and his father continued to hold Sherith Israel meetings to collect funds for land redemption, until the secret police arrested one of its members, who named Simon Velikovsky as the head of the group. This forced Simon into hiding and prompted my father to obtain visas through a friend for his parents to enter the Ukraine. He accepted responsibility for accompanying his parents safely out of Moscow, although he had to cease temporarily his medical studies. His brothers remained behind, Daniel to take care of his own family and Alexander to finish his chemistry studies at the Economic Institute and stay with the woman he loved, the younger sister of Daniel's wife, Genia.

Their exile to the Ukraine began on September 23, 1918 traveling by train and wagon. By the time they reached Poltava, where Simon's uncle Pavel would meet them, his mother had contracted dysentery. My father recalled from his childhood that his brother Daniel was treated for dysentery by a doctor who administered an infusion of white clover flowers, hence he collected the same herb, had it ground and administered it to his mother. Although several doctors said his mother would not survive, she eventually recovered. A year later, when Poltava began being terrorized by the Bolshevik forces, Daniel sent them money from Moscow so they could move to Kharkov, a city south of Moscow, where my father attempted to arrange safe passage to Palestine via Rostov and Caucasus. In doing so, he was taken by the secret police for questioning, but fortunately was only robbed and then released quickly. Due to this incident, his mother insisted they stay in Kharkov until travel was safer and my father registered at Kharkov University medical school. Meanwhile Daniel and Alexander started an oil shipment business in Moscow until Alexander was arrested for selling goods that were not properly registered. While Daniel managed to get his brother released, he himself spent a brief time in jail for trying to sell diamonds, and my father returned to Moscow for ten days to help get his brother released. By the time he arrived, Daniel had already been released, so my father decided to take advantage of the opportunity to have his credits transferred from Kharkov University to Moscow. He also obtained permission to take final medical exams and passed.

After the civil war was officially over in January, 1921, my father managed to obtain permission for him and his parents to leave Moscow for Germany. He also obtained documents to enable his brothers and their families to leave Russia, if they wished to leave. In the early summer of that year the brothers accompanied their parents to the frontier town on the German border where they would leave the country. My father obtained Lithuanian citizenship (because his mother was born in Grodno), returned to Moscow to graduate and obtained permission to leave Russia. He then went to Berlin to be with his parents and to start his post-doctoral residency at the Berlin Charity Clinic. Except for one more visit to Krovno to see Daniel, it was the last time he would ever see his brothers. Daniel was unsure whether he too should leave Russia and asked my father for advice. My father did not try to persuade him to leave because he knew that Daniel's wife Genia was a strong-willed Muscovite and would never leave.

Years later he blamed himself for not doing so. Alexander, with whom my father had been very close, never even considered leaving Russia, his studies and the woman he loved.

His brothers died in their eighties some years before my father's death. My father did not contact them for fear that they would be in trouble because of my father's escape from Russia. When he did send a message from the U.S. with a friend to one of his brothers, the brother told the caller that he did not have a brother. For years, my father and later my mother sent them clothing and money to Russia through France. It was understood that I would never meet my uncles. During détente my father talked of visiting his brothers, but fear that he would not be permitted to leave Russia kept him from going.

Right before Alexander's death in 1973, my father received a letter from him saying that he had followed my father's career through articles published in the Russian press.

While in Berlin with his parents, the idea of a Hebrew renaissance was still on my father's mind. He convinced his father to sponsor a scholarly journal that would contain works of Jewish scientists to develop the Hebrew language and to become the foundation of the future Hebrew University. My father then met with Heinrich Loewe, one of Germany's first Zionists, and told him of his ideas of co-editing this journal, to be titled *Scripta Universitatis atque Bibliothecae Hierosolymitanarum*. With Loewe's guidance *Scripta* became a reality. Chaim Weizmann gave his support and Albert Einstein became the editor of the mathematics/science section. After my grandfather gave my father the funds to start *Scripta*, my grandfather and my grandmother went to live in Palestine. My father stayed in Berlin where he and Loewe edited the papers and undertook the task of having them translated into Hebrew before publishing them in *Scripta*.

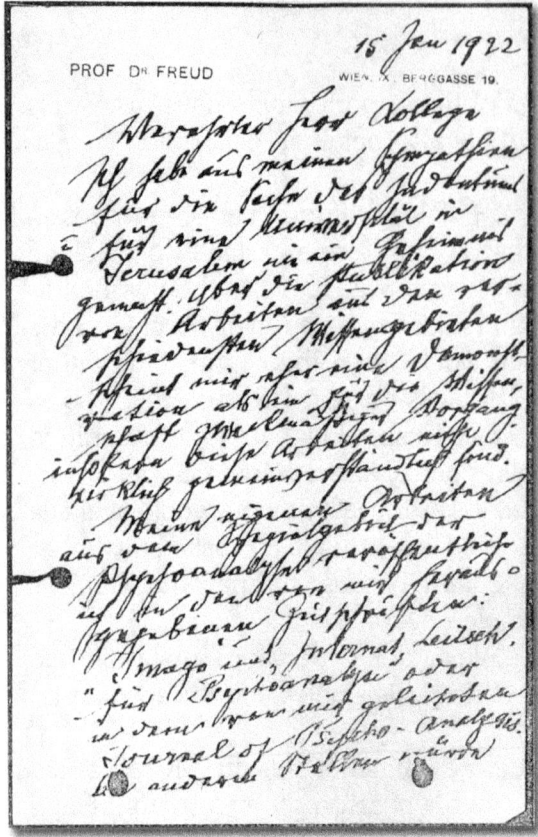
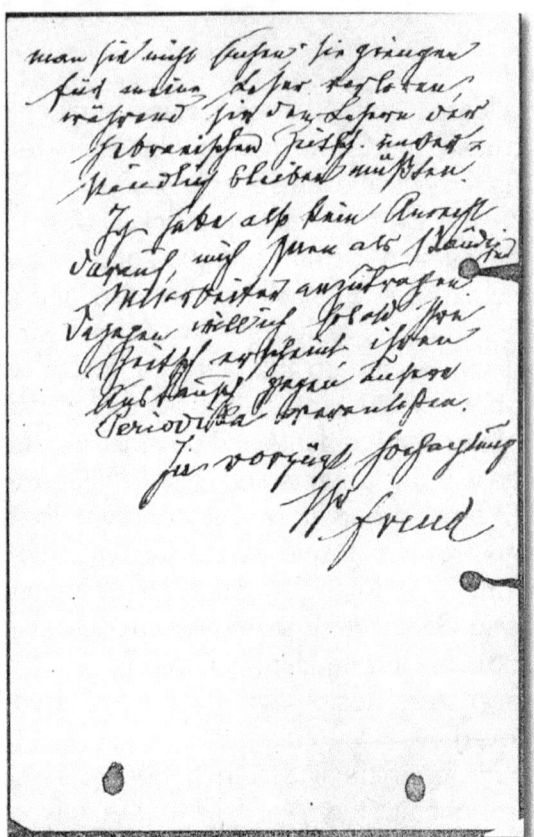

Verehrter Herr Kollege,

Ich habe aus meinen Sympathien für die Sache des Judentums und für eine Universität in Jerusalem nie ein Geheimnis gemacht. Aber die Publikation von Arbeiten aus den verschiedensten Wissensgebieten scheint mir eher eine Demonstration als ein für die Wissenschaft zweckmässiger Vorgang, insofern diese Arbeiten nicht wirklich gemeinverständlich sind.

Meine eigenen Arbeiten aus dem Spezialgebiet der Psychoanalyse veröffentliche ich in den von mir herausgegebenen Zeitschriften: "Imago" und "Internat. Zeitsch. für Psychoanalyse" oder in dem von mir geleiteten Journal of Psycho-Analysis. An anderen Stellen würde man sie nicht suchen. Sie gingen für meine Leser verloren, während sie den Lesern der hebräischen Zeitsch. unverständlich bleiben müssten.

Ich habe also kein Anrecht darauf, mich Ihnen als ständiger Mitarbeiter anzutragen.

Dagegen will ich sobald Ihre Zeitsch. erscheint, ihren Austausch gegen unsere Periodika veranlassen.

In vorzügl. Hochachtung
Ihr Freud

Honoured Colleague,

I have never made a secret out of my sympathies for the Jewish case and for a University in Jerusalem. But the publication of papers from various fields of knowledge seems to me more a demonstration than an act suitable for science, since these papers are not really commonly understandable.

I publish my own papers in the special field of psychoanalysis in the journals which I edit: "Imago" and "Internat. Zeitsch. für Psychoanalyse" or in the Journal of Psycho-Analysis which is supervised by me.

In other places nobody would look for them. They would get lost for my readers, whereas for the readers of the Hebrew journal they would remain incomprehensible.

So I have no right to offer myself to you as a contributor.

However, as soon as your journal appears, I will arrange an exchange with our periodicals.

Respectfully
Your Freud

The papers accepted for publication were authored by the most prominent Jewish scholars and scientists in the world. Freud, however, refrained from contributing (cf. his letter).

Soon after, in 1923, Chaim Weizmann asked my father, then 27 years old, if he would like to be the president of the new Hebrew University, but my father did not accept. The university was established in 1924 and inaugurated in 1925 when two volumes of *Scripta* were placed in front of Lord Balfour who came to the inauguration ceremony.

It was through his work with *Scripta* that my father met my mother, Elisheva Kramer, a violin student studying under the well-known concert violinist, Adolf Busch. My mother volunteered to help my father organize work on the journal. My father recalled that courting my mother during the years of famine following the first world war, my father visiting her studio asked her if she had any bread in the house. She looked in the cupboard and said "no", whereupon my father took out a loaf of bread from under his coat. He did the same thing with cheese and fruit and so on; my mother, each time looking in her cupboard. My mother gave her earnings to her sister Selma, who was pregnant and alone.

My parents became close and in the early spring of 1923 they were married on April 15th in Berlin. The next afternoon they spent "honeymooning" at the Berlin University Library, researching how scientific publications publish their works. Later this was to become their anniversary tradition for many years to come. My father often spoke of my mother's noble character, honesty and altruistic nature. Often, he said, she would walk a considerable distance across Berlin to visit a girl with a brain tumor and would play the violin for her. He considered my mother's presence in his life a blessing.

During the summer of 1923 they traveled to Italy by way of Egypt, and on October 30th arrived in Palestine to settle and raise a family. Although my father started to practice medicine there, a great deal of his energy went to straightening out his father's legal problems regarding Ruhama, the agricultural settlement he founded in the southern Negeb. The original Sherith Israel members had been forced off the land by Turks and the settlement owed its debtors 29,000 pounds. The problem was not resolved for seven years.

During that time my father felt responsible to help my grandfather achieve his vision of Zionism. Together they tried to start an Institute of Jewish Study in Tel-Aviv and to revive publication of *Scripta* which had been discontinued due to lack of funds.

My mother devoted her time to organizing the Palestine String Quartet. On February 1, 1925 my sister Shulamit was born in Jerusalem and I was born in Jerusalem the following year on April 26, 1926. Later that year my parents moved to the Carmel to a stone house overlooking the Mediterranean Sea.

Sabta and Saba

Sabta

with Daniel

Sabta and Saba

My father's mother, Sabta (Hebrew for grandmother), Biela Rachel Grodensky, the eldest child of a large family, was taken out of school to help her mother take care of her nine younger siblings. Despite her lack of formal education Sabta was a self-taught linguist speaking several languages fluently and was determined that her three sons receive the best education possible.

My father told me that when he was young, his mother would come into the boys' room in the middle of the night and complain about Aba's father. My father was attached to his mother and to the end of his life frequently sighed, "Oh, Mamitchka". His mother was instrumental in my father's relentlessly pursuing education. Her ambition for her son together with her narcissistic relationship with him brought about both his triumph and his vulnerability. Toward his father, he had unresolved anger – objecting to his father's treatment of his mother.

In March 1928, my father's mother died in Tel-Aviv.

My father described his mother's final moments in a letter to my mother to the Carmel.

Thursday

Dear Elisheva,

I just now spoke to you by telephone and told you how it is here. Be calm, strengthen yourself, occupy yourself with the children. I am with trust in God. Mother, as I told you, seems to have had a stroke. But she is not suffering and lies calmly with open eyes. She recognized me. Dear God helps.

Always Yours Emanuel

Sabta's grave

Saba

Many years later when we visited Israel, my son Rafael and I were given the assignment to assess the condition of my grandmother's grave. We were told it was surrounded by a wrought iron fence. We spent hours walking through isles of tombstones in the hot, cloudless August sun, reading inscriptions, never locating the site. It was later located by someone else and I wondered how that was possible since we searched for it systematically over and over again.

My father's mother seemed to have suffered from Alzheimer's and died of a stroke. Thus, Aba feared senility and did not tolerate forgetting even one name.

My grandfather (Hebrew: Saba), Simon Yehiel Velikovsky, devoted his life to three goals: to redeem the land of Israel, to establish a Jewish University there, and to restore Hebrew as a living language once again. He succeeded in accomplishing all three.

In *Ages in Chaos*, my father wrote about his father's scholarly achievement:

> He contributed to the revival of the language of the Bible and the development of modern Hebrew by publishing (with Dr. J. Klausner as editor) collective works on Hebrew philology, and to the revival of Jewish scientific thought by publishing, through his foundation, *Scripta Universitatis*, to which scientists of many countries contributed and thus laid the groundwork for Hebrew University at Jerusalem. He was the first to redeem the land in the Negeb, the home of the patriarchs, and he organized a co-operative settlement there which he called Ruhama; today it is the largest agricultural development in Northern Negeb.

My father's father at fifteen left home to attend the famous yeshiva of Volojin where he spent sixteen hours a day studying. Here he made friends with classmate, historian, Simon Dubnov, who later dedicated many articles to him in the newspaper Haaretz. But Simon proved to be too liberal for the traditional rabbinical teachings, and instead pursued a career as a business man, owning one of the most successful companies in Moscow that imported tea and other merchandise.

Saba was one of the leading members of the Zionist community, first in Vitebsk and later in Moscow and attended the Second Zionist Congress in Basel as a delegate where he met Theodore Herzel. Through this, he helped create the Anglo-Palestine Bank (later known as the Jewish Bank).

Dr. J. B. SAPIR

DER ZIONISMUS

Eine populär-wissenschaftliche Darlegung des
Wesens und der Geschichte der zionistischen
Bewegung

AUTORISIERTE ÜBERSETZUNG
VON
A. BENJAMIN

Vorliegendes Werk ist von der Commission der Wilnaer Zionisten
mit dem Preise von S. J. WELIKOWSKY
gekrönt worden.

BRÜNN
JÜDISCHER BUCH- UND KUNSTVERLAG
1903

Das vorliegende Buch verdankt sein Erscheinen der freigebigen Spende des Herrn S. J. Welikowsky, der den Wilnaer Zionisten 300 Rubel zur Verfügung stellte als Prämie für ein populäres Werk, welches in allgemein verständlicher Weise eine möglichst erschöpfende Uebersicht des Zionismus (seines Wesens, seiner Geschichte und Entwicklung, seiner Ziele und Bestrebungen) gibt. Von den zum Wettbewerbe eingereichten Arbeiten wurde das Werk des Herrn Dr. J. B. Sapir mit dem Motto: „Geduld und Ausdauer bringen alles fertig", als den obgenannten Forderungen am meisten entsprechend erkannt. Die Fülle des gesammelten Materials, die Leichtigkeit der Sprache, die möglichst übersichtliche Gruppierung der verschiedenen einschlägigen Detailfragen berechtigen zu der Hoffnung, dass die weite Verbreitung dieses Buches viel dazu beitragen wird, in der lesenden Masse richtige Ansichten über den Zionismus und sein Wesen einzubürgern. Auch gibt sich die Kommission der Hoffnung hin, dass die Arbeit Dr. Sapirs der verheissungsvolle Beginn von bedeutend umfassenderen, wissenschaftlich begründeten Forschungen über alle Fragen und theoretischen Grundlagen des Zionismus sein werde.

Die Kommission der Wilnaer Zionisten.

When the Wilnaer Zionist Committee needed money to print Joseph Sapir's *Zionism*, Saba donated his own money to do so. This fact is honored on the backcover of the book:

> The appearance of the present book is indebted to the generous donation by S.J. Velikovsky, who provided the Wilnauer Zionists with 300 Rubles as prize for a popular volume, that is to provide in a generally understandable way an overview of Zionism that is as exhaustive as possible (its nature, its history and development, and its aims and goals). Of the papers entered for the competition, Dr. J. B. Sapir's work with the motto: "Patience and persistence can accomplish everything" was the one acknowledged to most effectively satisfy the above conditions. The large amount of collected materials, the facility of the language, and the good overview that the groupings of detailed questions provides, allow the hope that the wide circulation of this book will help install in the reading masses proper views concerning Zionism and it's nature. The commission also hopes that the work of Dr. Sapir will be the promising start of more comprehensive and scientifically based research concerning all the questions and theoretical bases of Zionism
>
> The Commission of the Wilnaer Zionists.

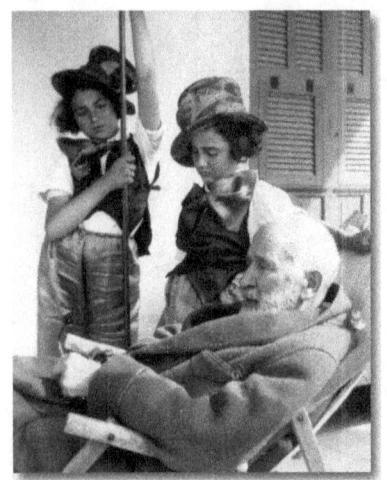

Saba's house

Saba's grave

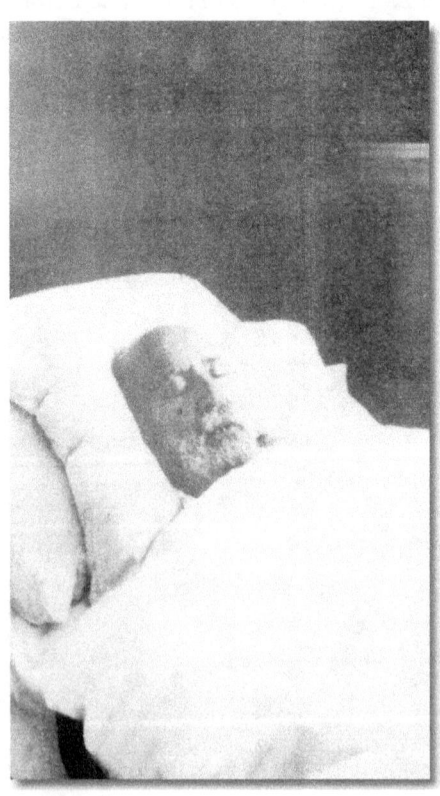

Saba

I remember walking next to my father on the streets of Tel-Aviv. It was the Sabbath and I was ten. I ran to keep up with his stride. My father seemed not to be concerned with me, for he was concerned with his father. There was urgency in his pace. Saba, my paternal grandfather, was sickly with diabetes. He also suffered incessantly from an eyelash growing inside his eyelid.

Ruth, Saba and Shulamit on Purim

He rested in a dimly lit room elaborately furnished with ostentatious, white ivory furniture. My father's devotion to his parents was complete. His as yet unpublished autobiography tells of his escape from Russia with the purpose of getting his parents out of the country. The autobiography tells of his life until the 1940's when *Stargazers and Gravediggers* picks up the story. Just as my father worked to vindicate his father, finding himself distracted from his own life and family, so I spent much of my time trying to help my father during years when the fury of the scientific establishment weakened his body and spirit.

I went to visit my grandfather because he loved me and because it was good to walk alongside my father and to climb the ledge of my grandfather's house. Agile, I did somersaults on the bars of the windows, climbed between the bars, rode a bicycle and rolled toward the Tel-Aviv beach on roller skates.

My father's father adored me. I sat on his lap on the porch of his blue stucco house in Tel-Aviv as he said "Rihamti et Ruhama." ("I pity Ruhama").

My middle name, Ruhama, was in honor of the southern land in Israel that my grandfather had purchased and donated to the Jewish National Fund. Saba's fund raising for the purchase of the land led to litigation and aggravation for both my grandfather and father. Although vindicated, Saba, a man of integrity, never recovered.

He had a new wife who was glamorous and did not like to take care of him. When her sons learned there would be no inheritance, she disappeared and my grandfather bemoaned the loss. I suspect that my father disliked her and "helped" her leave, although I have no such knowledge.

My father's father knew his youngest son was destined to become a man of distinction. On his death bed he bid my father to save the world having no way of knowing what an immense contribution my father would make toward the clarification of our past history, its implication for the future of our world, and the suffering my father was to experience.

My grandfather respected my father. This son helped him escape from Russia. Peering over his dark glasses to study my father's expression which betrayed his thoughts, Saba knew he was about to die and give my father freedom to travel out of Palestine and onto new vistas of greatness.

Saba's Mother
Sara Hotimsker Velikovsky

Saba's Father
Jacob Velikovsky

Saba

Sabta and Saba

It was 1937. My grandfather was buried in a cemetery in the middle of Tel-Aviv (the same one where Sabta had been buried) in a section where great men of distinction were interned, men such as Malik. Saba, having been an ardent zionist donating land he had purchased with his own funds to the Jewish National Fund, was buried among great men. Bialik's grave is next to his. My father was informed that the burial site was agreed to by the cemetery directors because of Saba's wealth, not because of his devotion to the Jewish cause. This criteria angered my father sufficiently that he withheld payment of the bill for some time. Eventually, he paid the bill, but felt guilty and kept sending monetary gifts to the cemetery to the end of his life. My mother repeatedly explained his motivation in paying the bill late. The delay, she explained, was a statement – not a delinquency.

Saba's mother, Sara Hotimsker, was a descendant of two great codifiers of Jewish law, Ezra the Scribe, and Joseph Ben Ephraim Caro. And Sara's father, Jacob Hotimsker, a dayan, or religious judge, was revered as a holy man in his community of Mstislar.

Saba's father, Jacob Velikovsky, valued scholarly pursuits. Before Hebrew became a spoken language, Jacob diligently perfected his Hebrew, making it the only language he spoke on the Sabbath.

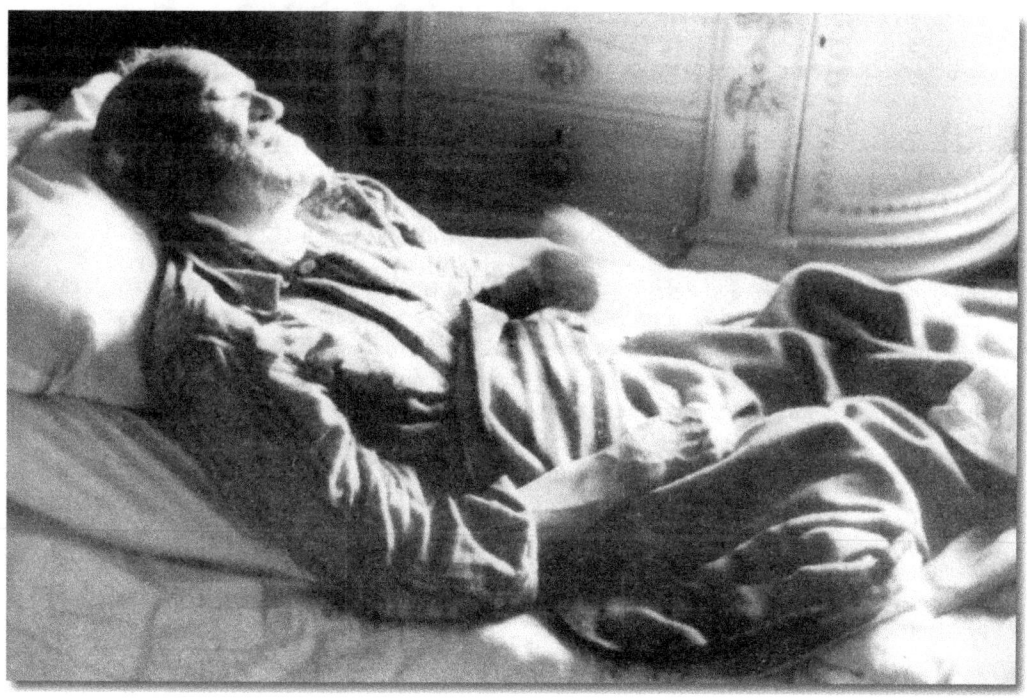

Ima, Aba, Sabta, Shulamit and Saba
Jerusalem 1925

Ima

Ima

Aba, Shulamit, Ruth and Ima

Ima

My mother (Hebrew: Ima) was born in Hamburg on July 27, 1895. She was one of six siblings. Her mother had two stillbirths (both boys). The only boy to survive was Sigfried ("Sigi"), a twin of Trudel.

As a little girl my mother ate only when she was paid. The coins were ceremoniously placed in a box by her father and saved for a present for her mother, who for years had been bed ridden with diabetes.

After my mother experienced famine during World War I, her eating habits changed and she began to gain weight. She especially gained weight after childbirth and she often said she envied people who ate whatever they wanted without gaining weight, and added, she also envied people who believed in the afterlife.

My mother told me that when she was young, she had been very slender – so slender that she wore a brooch to differentiate her front from her back She was beautiful. Many men wanted to marry her. She knew she was destined to marry a great man and waited until my father appeared. Shazar, who later became president of Israel, wanted to marry her. He embraced and kissed her in front of dignitaries on my parents' visit to Israel in the 1970's. My father was happy for her. He often told me I had an unusual mother.

Aba, Ima and President of Israel

Birthdays, as my mother told it, were memorable meaningful celebrations in her home. One year my mother was a "bad" child and therefore the customary birthday table had only a whip on it. My mother was then escorted to another room where many presents covered a table. During another birthday, different in spirit, my mother's mother intentionally spilled a glass of chocolate milk on the tablecloth pronouncing the children could relax and have a good time. This impressed my mother, who cared more about people than about things. She never got upset when something broke or was lost. My father, on the other hand, became angered when something borrowed was not returned. He looked after his belongings.

Ima's Mother
Fanny, née Schur

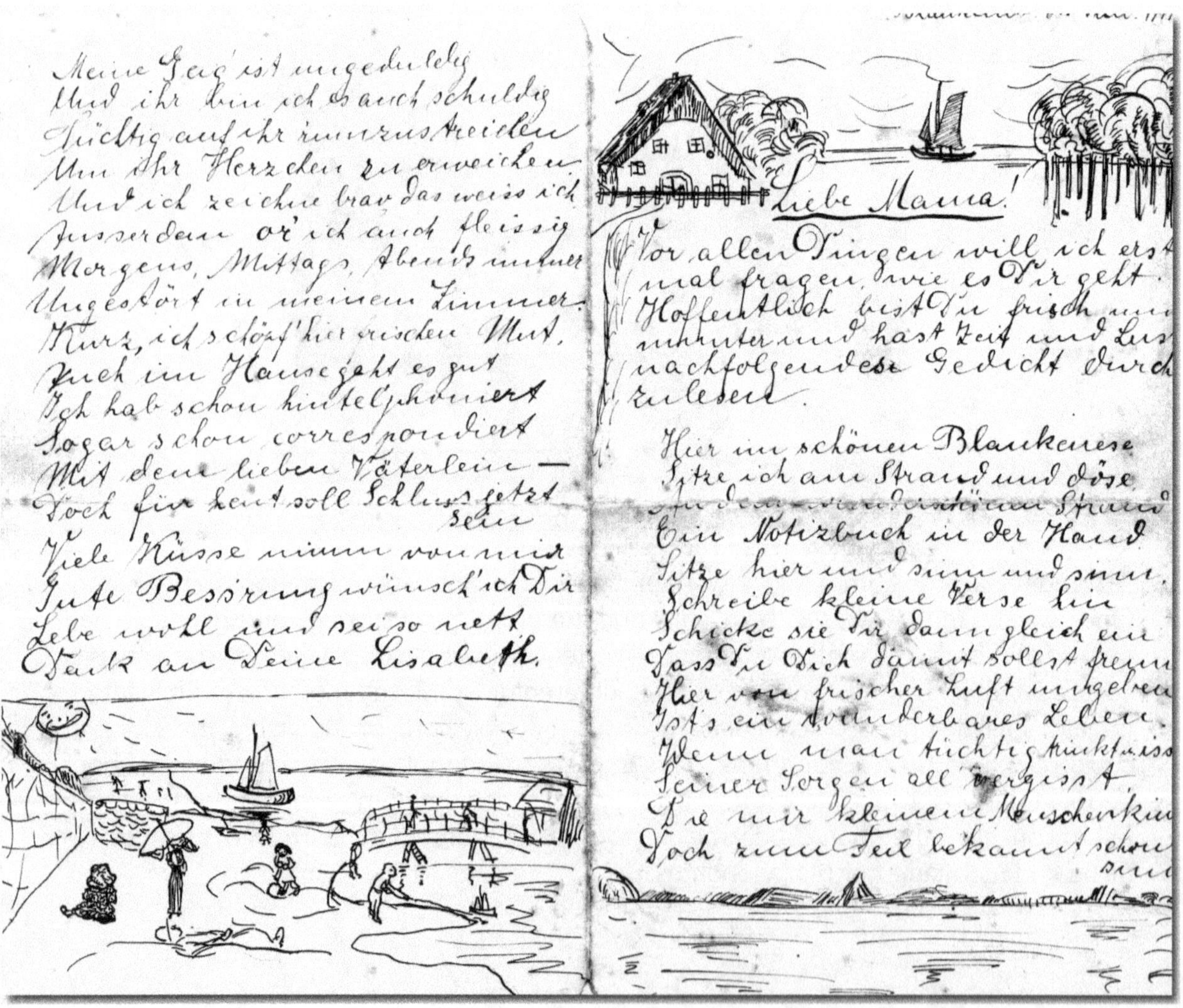

For many years my mother's mother suffered with a severe case of diabetes and was confined to bed. She died when my mother was eighteen. My mother at the age of sixteen, having moved from home to study the violin, wrote her mother several poems.

Blankenese, 7 July 1911	Blankenese, 7. July 1911
Dear Mother,	Liebe Mama!
Before everything I want to ask you how you are. Hopefully you are fresh and cheerful and have time and desire to read the following poem:	Vor allen Dingen will ich erst mal fragen, wie es Dir geht. Hoffentlich bist Du frisch und munter und hast Zeit und Lust nachfolgendes Gedicht durchzulesen:
Here in beautiful Blankenese I sit on the beach dosing On the wonderful beach, With a notebook in my hand, I sit here and think and ponder Writing down little verses. I'll send them to you straight away So that you can enjoy them. Here, surrounded by fresh air, It is a wonderful life. If you eat and drink well You will forget all your worries That to me, little soul, in part, are already known It leaves father without peace Whether I eat well. So I wanted to tell you That my nasty stomach Consumes terribly much these days, Everything that is brought to it. My appetite is not normal In the mornings I eat two eggs In the evenings I eat two again And drink lots of milk with them. Sometimes I also eat cheese Or fish with mayonnaise Often, there is fruit afterwards Which my stomach loves dearly. Second breakfast is taken at the beach Often with some sand mixed in. And then around 1 pm There is something good for dinner: There is soup, there is roast, There's vegetables along with salads. Around 6 there is coffee	Hier im schönen Blankenese Sitze ich am Strand und döse, An dem wunderschönen Strand Ein Notizbuch in der Hand. Sitze hier und sinn und sinn, Schreibe kleine Verse hin Schicke sie Dir dann gleich ein Dass Du Dich damit sollst freun. Hier von frischer Luft umgeben, Ist's ein wunderbares Leben. Wenn man tüchtig trinkt und isst, Seiner Sorgen all' vergisst, Die mir kleinen Menschenkind Doch zum Teil bekannt schon sind. Vatern lässt es keine Ruh Ob ich auch gut essen tu. Darum wollt' ich Dir es sagen, Dass mein unverschämter Magen Furchtbar viel zur Zeit verschlingt, Alles, was man ihm nur bringt. Mein App'tit ist nicht geheuer. Morgens esse ich zwei Eier, Abends ess' ich wieder zwei Und trink tüchtig Milch dabei. Manchmal esse ich auch Käse Oder Fisch mit Mayonnaise Oft gibt es auch Frucht nachher, Was mein Magen liebet sehr. Zweites Frühstück gibt's am Strand Oftmals untermischt mit Sand. Und so ungefähr um eins Gibt zu Mittag es was Feins: Gibt es Suppe, gibt es Braten Gibt's Gemüse, nebst Salaten Und um sechs gibt es Kaffee

Meine liebe Mama!

Ach, wie hab ich mich gefreut
Als am Donnerstag, zur Zeit
Da ich grad zum Essen ging
Deine Zeilen ich empfing.
Vielen, vielen Dank dafür.
Grosse Freude macht es mir.
Jetzt, Mama, mach ich Dir klar,
Dass ich ganz begeistert war
Als ich Montag ging zum Essen.
Und die Taja, wie besessen
Ganz glückstrahlend zu mir lief
Und ganz laut und freudig rief:
"Ihre Mutter schrieb an mich
Eine Karte nachträglich!
Und dazu noch solch eine feine
Mit 'nem Briefkasten, so eine!"

Or, if you prefer, also tea.
But with all that abundant food
The spirit isn't forgotten either.
My violin is impatient
And I owe it to her
To bow up and down on her like an addict
To soften her heart.
And I sketch regularly, I know.
Furthermore I prey diligently
In the morning, noon and evening
Unbothered in my room.
In short, I gather new courage here,
All is well in the house.
I have already phoned
Even corresponded
With my nice little daddy –
But enough for today
Many kisses take from me
I wish you a good recuperation,
Farewell and be so kind
To think of your Lisabeth.

My Dear Mama,

Oh, how happy I was,
When, on Thursday, just
As I was walking to dinner,
I received your lines.
Many, many thanks for it.
It gave me great joy.
Now, Mama I am going to tell you
That I was all excited
When I was going to dinner on Monday
And Daja, almost out of her mind
Came running to me, full radiant with bliss
And called out loud and with joy:
"Your mother wrote me
A postcard after the event!
And what a nice one,
Such a one with a mailbox on it."
Yes, Daja it got happy
And me it got totally charmed.
And Mrs. Israel no less,
Even Papa and the children.

Little girls, all different
Olga, Ida, Frieda, Katie
Are all very well acquainted with me,
This friendship started at the beach.

Oder, wenn man will, auch Tee.
Doch bei all dem vielen Essen
Wird der Geist auch nicht vergessen.
Meine Geig' ist ungeduldig
Und ihr bin ich es auch schuldig
Süchtig auf ihr 'rumzustreichen
Um ihr Herzchen zu erweichen.
Und ich zeichne brav, das weiss ich,
Ausserdem or' ich auch fleissig
Morgens, Mittags, Abends immer
Ungestört in meinem Zimmer.
Kurz, ich schöpf' hier frischen Mut,
Auch im Hause geht es gut.
Ich hab schon hintel'phoniert
Sogar schon correspondiert
Mit dem lieben Väterlein –
Doch für heut soll Schluss jetzt sein
Viele Küsse nimm von mir
Gute Bess'rung wünsch ich Dir
Lebe wohl und sei so nett
Denk an Deine Lisabeth.

Meine liebe Mama,

Ach wie hab' ich mich gefreut
Als am Donnerstag, zur Zeit
Da ich grad zum Essen ging,
Deine Zeilen ich empfing.
Vielen, vielen Dank dafür.
Grosse Freude macht' es mir.
Jetzt, Mama, mach ich Dir klar,
Dass ich ganz begeistert war,
Als ich Montag ging zum Essen
Und die Daja, wie besessen
Ganz glückstrahlend zu mir lief
Und ganz laut und freudig rief:
"Ihre Mutter schrieb an mich
Eine Karte nachträglich!
Und dazu noch solch ne feine
Mit 'nem Briefkasten, so eine!"
Ja, die Daja hat's beglückt.
Mich, mich hat es ganz entzückt.
Und Frau Israel nicht minder,
Selbst Papa und seine Kinder.

Ganz verschied'ne kleine Mädchen
Olga, Ida, Frieda, Kätchen
Sind mit mir intim bekannt,
Diese Freundschaft ist vom Strand.

Ja, die Daga hat's beglückt
Mich, mich hat es ganz entzückt.
Und Frau Israel nicht minder,
Selbst Papa und seine Kinder.

Ganz verschied'ne kleine Mädchen
Olga, Ida, Frieda, Käthchen
Sind mit mir intim bekannt.
Diese Freundschaft ist vom Strand.
Diese 3, 4, 5, 6 Mädchen.
Olga, Ida, Frieda, Käthchen
Sind wohl 3, 4, 5, 6 Jahre,
Haben alle helle Haare,
Laufen alle ohne Schuh
Sprechen falsches Deutsch dazu
Wie die Freundschaft ist gekommen
Sei aus folgendem entnommen
Neulich sass ich an dem Strand
Zeichnete dort allerhand
Zeichnete dort dies und das
Auch mein vis-à-vis zum Spass.
Und da kam die kleine Ida
"Himmel" rief sie „das ist die da"
"Zeichnen Sie mir auch mal, mich!"
"Ja, mein Kind, denn setze Dich.
4mal hab ich sie gezeichnet.
Denn sie schien mir sehr geeignet
Bald da kam ein andres Mädchen
Hiess sie Olga oder Käthchen?

Warum zeichnen's mir denn nicht?
„Gern mein Kind, denn setze dich
Als ich nun genug „gemalen"
Musste ich sie auch bezahlen
Jeder kriegt ein Kupferstück,
Kinder, war das für ein Glück!
Sagt, wo tut ihr das denn hin
„Das kommt in den Sparstrumpf rin"
Andern Tages früh am Morgen
Ging vergnügt und ohne Sorgen
Ich gemütlich an den Strand
Doch wer gibt mir da die Hand
Olga, Ida, diese beiden
Möchten mich wohl gerne leiden
Brachten mit 10 andre Gören
Musste alle Namen hören;
Doch sie alle zu behalten
Traue ich meinem lieben, alten
Bald schon 61jährgen Schädel
Nicht mehr zu ich altes Mädel
Alle, alle musst ich malen
Keinen einzigen bezahlen
Denn mein denk Dir bloss
War vollständig inhaltslos.
Nun für heute genug
Dieser Brief trefft Dich vergnügt
Hoffentlich erfreut er Dich
Das wünscht Lisbeth aufrichtig.

These 3, 4, 5, 6 girls Olga, Ida, Frieda, Katie Are all probably aged 3, 4, 5, 6, years, They all have blonde hair They are all going without shoes And on top, speak funny German. How this friendship came about May be deduced out of the following. The other day I was sitting on the beach Drawing all kinds of things Drawing there, this and that And also, for fun, the one sitting across from me. And there came little Ida strolling. "Hurry" she cried out, "It's this one! Also, draw me, couldn't you?" "Yes my child, sit down!" Four times I drew her For she seemed well appropriate to me. Soon there came a young lady, Was she named Olga, or Katie? "Why don't you draw me?" "Sure, my dear, sit down near me." Now that I had been painting enough, I needed to pay them accordingly. Each received a piece of copper A piece of luck for them, "A whopper." "Say where you're going to put it" "It will go to be kept in savings." The next day early in the morning I went, with happy heart and free of worries To the beach. But who greets me with a handshake? Olga, Ida, those two Must have thought of me with fondness, Brought along 10 other sweethearts. Each one's name was told to me Yet to keep them all in mind Is too much for my dear old skull Going on sixteen, Me poor old maiden. All, but I had to paint None at all I needed to pay For my wallet, just think how clever, Showed no content whatsoever. Now for today this letter's sufficient. May it find you in a bright mood. With hopes that it will give you pleasure This is what Lisebeth wishes from her heart.	Diese 3, 4, 5, 6 Mädchen Olga, Ida, Frieda, Kätchen Sind wohl 3, 4, 5, 6 Jahre, Haben alle helle Haare, Laufen alle ohne Schuh, sprechen falsches Deutsch dazu. Wie die Freunschaft ist gekommen Sei aus folgendem entnommen. Neulich sass ich an dem Strand zeichnete dort allerhand, zeichnete dort dies und das Auch mein vis-à-vis zum Spass. Und da kam die kleine Ida. "Himmel," rief sie, "das ist die da! Zeichnen Sie mir auch mal mich!" "Ja, mein Kind, denn setze Dich." 4mal hab' ich sie gezeichnet, Denn Sie schien mir sehr geeignet. Bald, da kam ein andres Mädchen, Heisst sie Olga oder Kätchen? "Warum zeichnens' mir denn nicht?" "Gern mein Kind, denn setze dich." Als ich nun genug "gemalen" Musste ich sie auch bezahlen. Jeder kriegt ein Kupferstück Kinder, war das für ein Glück! "Sagt, wo tut ihr das denn hin." "Das kommt in den Spartop rin." Andern Tages früh am Morgen Ging vergnügt und ohne Sorgen Ich gemütlich an den Strand. Doch wer gibt mir da die Hand? Olga, Ida, diese beiden Mochten mich wohl gerne leiden, Brachten mit 10 andre Gören, Musste alle Namen hören. Doch sie alle zu behalten Trau ich meinem lieben alten Bald schon 16jährgen Schädel Nicht mehr zu, ich altes Mädel. Alle, alle musst ich malen, Keinen einzigen bezahlen, Denn mein Beutel, denk Dir bloss, War vollständig inhaltslos. Nun, für heute es genügt. Dieser Brief treff Dich vergnügt. Hoffentlich erfreut er Dich Das wünscht Lisbeth aufrichtig.

Ima's Father
George Kramer

Während unseres dortigen Aufenthaltes, wurde uns am 11. (elften) März 1862 (achtzehnhundertzweiundsechzig) zu Pittsburg ein Sohn geboren, welcher den Namen George Kramer erhielt. Wegen der Unruhen des Bürgerkrieges sahen wir uns schon bald nach der Geburt des Kindes genötigt, Amerika wieder zu verlassen und nach Rußland zurückzukehren, ohne daß die Anmeldung des Kindes bei dem Standesbeamten erfolgt wäre. Die Geburt des Kindes hat deshalb bis heute nicht angemeldet werden können. Ich bemerke noch, daß mir die bezüglichen amerikanischen Verhältnisse unbekannt waren und noch unbekannt sind und ich nicht weiß, noch wußte, ob und bei wem die Geburt eines Kindes überhaupt anzumelden werden müsse.

Hierüber

My mother's father, George Kramer, was born in the US, which I recently discovered through a legal document, part of which is duplicated on the adjacent page.
Its full translation reads:

Official Deposition

3rd June 1897

Before me, the official and sworn in attorney of law of Hamburg, H. L. W. Asher appeared:

Mr. Zemach Kramer from Tuckum in Kurland (East Prussia) currently present here working at Harient Street No. 14/15, legitimized by and declared:

In 1860 I emigrated with my wife Marianne, née Rosenthal, from Russia to the US of North America in order to secure my livelihood there. During our stay there, a son was born on the 11, March, 1862 in Pittsburgh to us who received the name George. Because of the unrest of the Civil War, we saw ourselves pressured to leave America shortly after the birth of the child and to return to Russia without having registered the child at the town Hall. For that reason we have been unable to register the birth of the child until now. I note that the relevant American circumstances were unknown to me then and are still unknown today and that I do not know, or knew, whether and where the registration of the birth of a child ought to be undertaken.

About this, this protocol was taken down in original writing and will remain in my, the Public Notary's safekeeping. After due reading and acknowledgement by Mr. Longerentun and myself the notary, it was signed by myself the notary, under affixation of my seal.
This happened in Hamburg on June 8.

Personal signature of Mr. Zemach Kramer

After the death of his wife, George Kramer married Mallie, a slender wigged religious young woman. Following a correspondence over several years with his children, Selma and Sigi in the US., my mother, Dorele, Trudel and Miriam in Israel, urging their father and Mallie to leave Germany, my grandfather, George Kramer closed his Jewish bookstore in Hamburg and emigrated with Mallie to Palestine.

I recall my grandfather squatting down in our Tel-Aviv apartment and teaching Shulamit and me prayers.

A few months my grandfather died of a heart attack; Mallie died a few years later.

Ima's Parents

Ima's Father

George Kramer, Export-Buchhandlung
Hamburg 13, Grindelallee 115   5/8 31

Mein teuerstes Liesel-heu'
6 Tage in Aw * den 27. Juli
hattest Du Geburtstag
* ich habe versäumt
Dir zu gratulieren. Ich
habe immer auf Brief
von
gewartet ist aber nicht
gekommen. Nur gestern
war von Dir ein dicker
Brief an Miriam auf
meine Adresse & sie
hat ihn spät abends
holen lassen & heute
habe ich sie noch
nicht zu sehen bekommen,
möglich, dass drin
auch ein zweiten

One of the hundreds of communications from my mother's father to my mother:

5.8.31

George Kramer

My Dearest Lieselchen,
6 days in Av and on the 27of July was your birthday and I failed to congratulate you. I was always waiting for a fetter from you but it never arrived. Just yesterday there was a thick letter from you to Miriam care of my address. She had it retrieved last night but I have not seen her yet today; perhaps there was a letter in there for me as well. Well, I don't want to hold you up any longer and want to send you – in Mallie's name, as well – our most heart felt Mazel-tov wishes for your birthday and best wishes for you and your nice husband and lovely children. Also, give greetings to Dorele and your husband.

Your Papa

From her early childhood my mother wanted to play the violin. She received a violin as a birthday "surprise", but her sister, Selma, pre-empted the surprise.

From Ima I inherited a deep love of classical music. She often said she was glad to have lived after Mozart. When she was at home, especially in my early years, she often played her violin. Music and my mother were synonymous. However, I did not like chamber music. Her playing chamber music meant that there were three other musicians in the house – an interference and intrusion into my time with my mother. When she played alone, I could stand next to her, write loving words on her music book and be close.

I hid from my piano teacher who came to our home. Placing a round coin in the palm of his hand, my father tried to impress me with the cost of the lesson. My viola teacher in the United States overdosed me on scales. Thereafter, except for playing chamber music with my mother, my musical career was over.

My mother, an exceptional teacher, made chamber music fun. As we got older, my sister played the cello and I the viola, and we left out the second violin part.

Ima and Selma
Hamburg, 1914

Ima and Selma (and Boris)

Shulamit and Ruth

Ima and Ruth

Ruth on the violin

My mother often travelled in Palestine with her chamber music group and left me and my sister in the care of her older sister, Miriam. Mimi had no children and lavished both love and food on us. She later took charge of my father's Tel-Aviv property when we sailed to the United States and packed up our apartment, putting everything in storage.

Ima's Quartet

Selma, her other sister, became a concert pianist.

My mother was infinitely patient when it came to studying the violin and to teaching music. Some of her violin pupils – like Itzhak Gitteles – became world renowned and introduced her at his concerts. She enjoyed the attention.

My father consistently encouraged my mother to arrange chamber music ensembles. He often sat close by and read his manuscripts while savoring the music. The musicians liked his presence, often playing pieces he requested. Music was my mother's talent, perhaps the only one Aba did not have (or at least did not cultivate).

To the end of her active life, she played chamber music, sometimes commenting that should her musicians know her age, they would not play "with such an old woman." She was eighty-seven. A soft spoken woman who never raised her voice to me, or to anyone, when leading a chamber group she became directive, even commanding and in charge. A few years before his death, my father bought her an Amati violin. Her slow vibrato and pleading tone hovered over every piece she played.

Ima
in Tel-Aviv

Itzhak Gittelis

Ima with
Arthur Schnabel

JERUSALEM MUSICAL SOCIETY

SEASON 1931-32.

SECOND CONCERT Saturday November 7th 1931

PROGRAMME

1. Allegro spiritoso and Siciliano
   for String Quartet - - - DALAYRAC

2. Quartet in D Minor "*Death and the Maiden*" - SCHUBERT
   Allegro — Andante con Variazioni — Scherzo — Presto.

3. Quartet in E flat major "*The Harp*" Op. 74 - BEETHOVEN
   Poco Adagio — Allegro. Adagio ma non troppo.
   Presto. Allegretto con Variazioni.

THE JERUSALEM STRING QUARTET

Violins — MRS. ELIZABETH WELIKOWSKY
         MISS MARGERY BENTWICH

Viola — MISS JENNY SCHMERZLER

Cello — MRS. THELMA YELLIN

NEXT CONCERT Saturday, November 21st 1931

Pianoforte Trios — Rameau, Brahms and Ravel

Living in Palestine

Ima and Aba

Aba, Ima and Shulamit

Living in Palestine

In Palestine we first lived on the Carmel. My father practiced as a physician.

Later he decided to seek further qualifications and trained to become a psychoanalyst in Vienna under Wilhelm Stekel. My parents travelled to Vienna several summers to study psychoanalysis. In the 1930's, after moving to Tel-Aviv, Aba opened the only psychoanalytic practice in Palestine.

He worked from early morning to late at night, my mother bringing his dinner on a tray. Occasionally, he would fall asleep and the patients tiptoed out of the office.

My mother was active in a string quartet and traveled around Palestine giving performances.

The care of my sister Shulamit and myself was given to housekeepers, yet my mother's love always came through. Esther, irreconcilable when her newly acquired passport was collected on a summer boat trip to Cyprus, cried like a baby instead of taking care of us. Etta ingratiated herself to my puppy, Hushi, who danced around her hoping for bones; and Chaya, a young, seductive Yemenite woman, never quite closed the bathroom door.

The Spanish Synagogue across from our Shadal Street apartment in Tel-Aviv had a big cherry tree. The leaves provided food for my silk worms. From egg dots in my shoebox, to fat worms, cocoons, butterflies and again little dotted eggs, I watched the cycle again and again.

My companion was Hushi, a police dog. I had selected the weakest puppy from a puppy litter of a dog belonging to one of my father's patients. She had had to grow some before I could take her home. When we moved to another location in Tel-Aviv, shortly before leaving for the U.S., Hushi disappeared. I found her chained in a former neighbor's yard – my parents had given her away. I abducted Hushi, and for hours in secret pulled off pea-like creatures of all sizes from her fur.

My cat, Mitzi, had disappeared while we were on a summer vacation. She was never found although for weeks I called "Mitzi" through the streets of Tel Aviv. In spite of my reluctance, we left Tel-Aviv beginning the journey that would end with our locating in New York. In Cyprus, I learned my rabbits had been killed in their cages by dogs. I felt responsible for the carnage. The final break with Palestine came for me when my parents were informed by Mimi that Hushi had been run over by a car.

Ruth and Mitzi

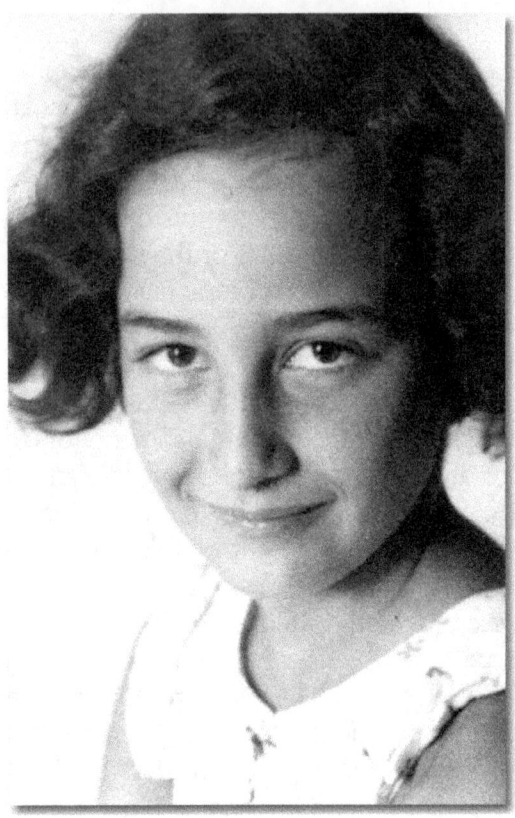
Ruth

Shulamit

My father called me "Rutinka", a Russian endearing affix; "Rut" when he was angry. I was "Ruthi" to my mother when I was good, and "Rut" when she felt neglected. To my sister, Shulamit, I was "Ruti" until recently.

Between themselves my parents shared a special relationship. When affectionate, he addressed her as "Shevic"; when angry – "Elisheva"; when business-like, he called her "Ima". To me, she was "Imale", and in the United States and until the end of her life, she was "Mommie". She called him "Monia", short for "Immanuel".

Between themselves, my parents mostly spoke German. Aba spoke Hebrew to my mother more often than she spoke Hebrew to him. In the United States, in one sentence they sometimes combined Hebrew, German, and English words. I spoke English to them soon after arriving in the United States. If they spoke in Hebrew to me, it meant my father was angry at me, he had just spoken in Hebrew to someone else, or, while depressed, was concerned about conversing on the telephone. My father rarely spoke Russian at home, yet the sound always reminds me of him.

My father was clean shaven, except in his thirties, when he grew a beard – his mother objected on the ground that his looking older made her appear older. Toward the end of his life he once again grew a mustache which my mother disliked. Always clean, although often walking barefoot, he took several showers daily, mostly cold, and when sickly with an irregular heart beat, soaked in a tub, claiming that a body submerged in water makes the heart work easier.

Aba and Shulamit

Living in Palestine

Ima's forte was clearly music and art. Her sculptures were particularly sensitive – but in art, Aba also had talent. He sketched and doodled dozens of profiles of men nurturing a variety of large noses and retreated chins – caricatures drawn with fine lines, men of character.

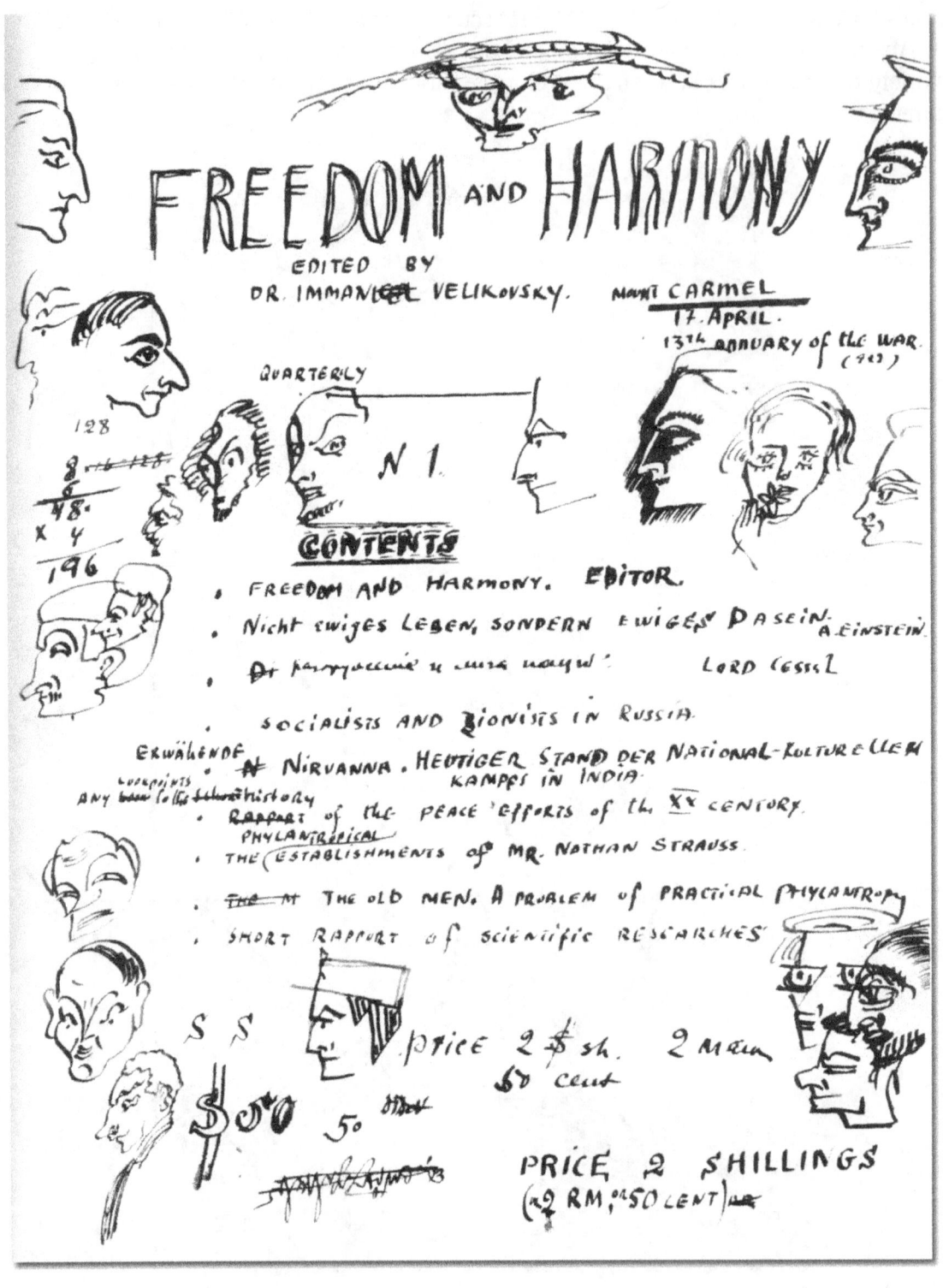

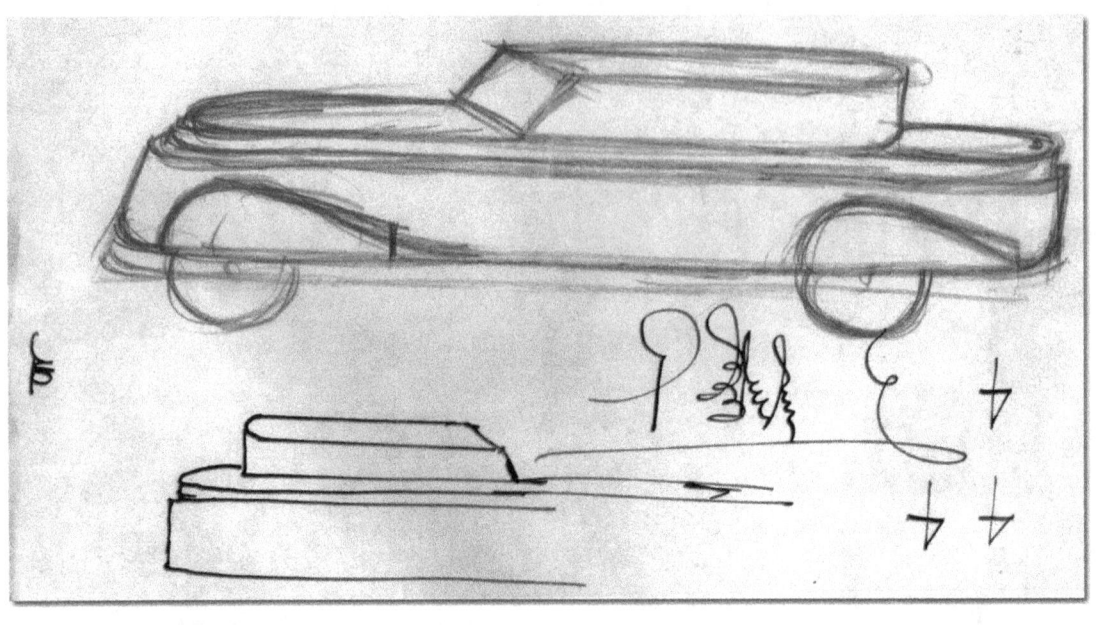

Tel-Aviv building before it was torn down

Living in Palestine

Aba also designed streamlined automobiles and was the architect for his Tel-Aviv building, which drew acclaim.

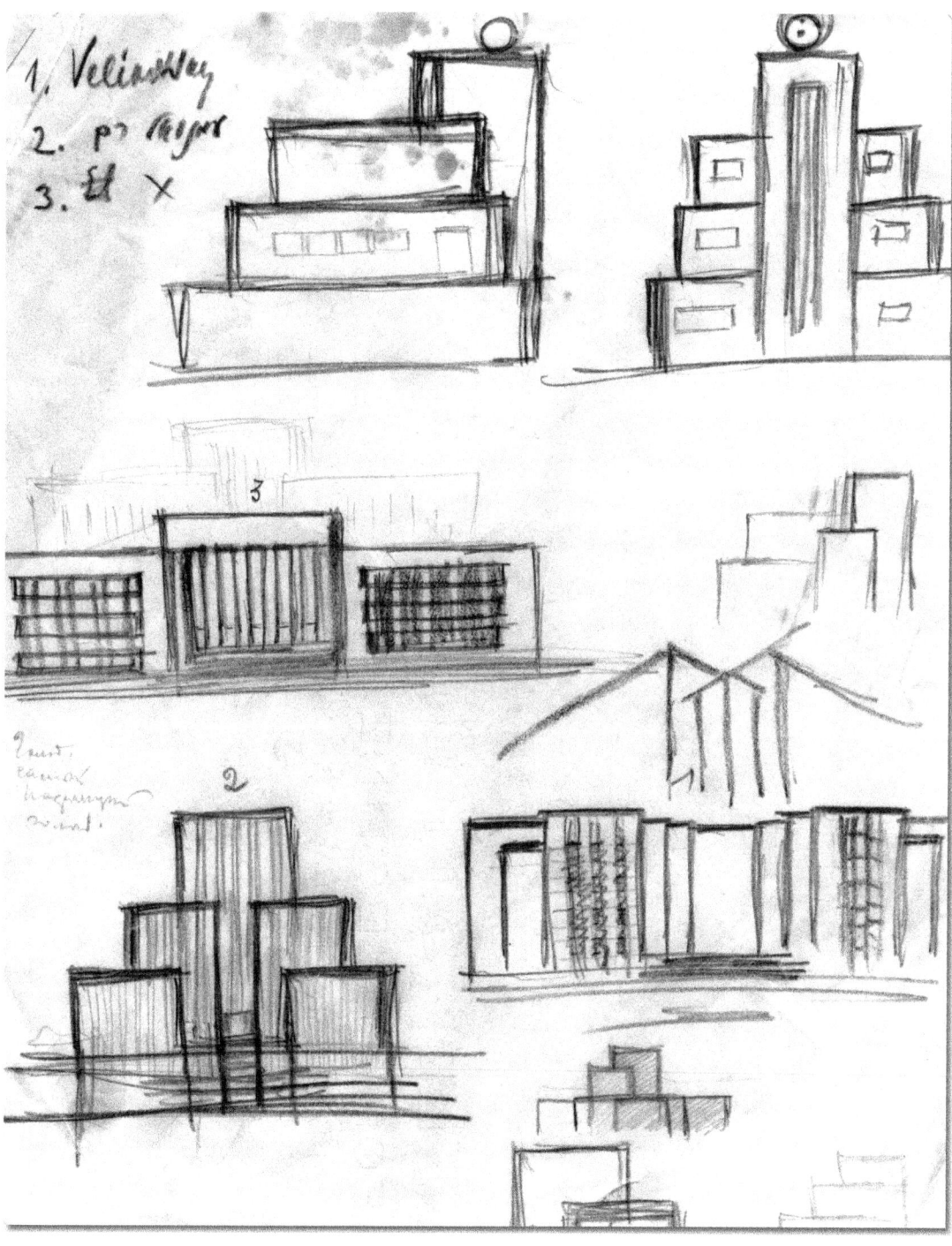

Aba and Psychoanalysis

Dr. C. G. Jung Küsnacht-Zürich
Seestrasse 228 28.3.30.

Sehr geehrter Herr Doktor,

Leider bin ich Mitte April nicht in Zürich. Ich werde erst Anfang Mai wieder anfangen. Dann kann ich Ihnen die Möglichkeit einer Ausbildung geben.

Mit vorzüglicher Hochachtung

Ihr ergebener

C. G. Jung

Letter 1

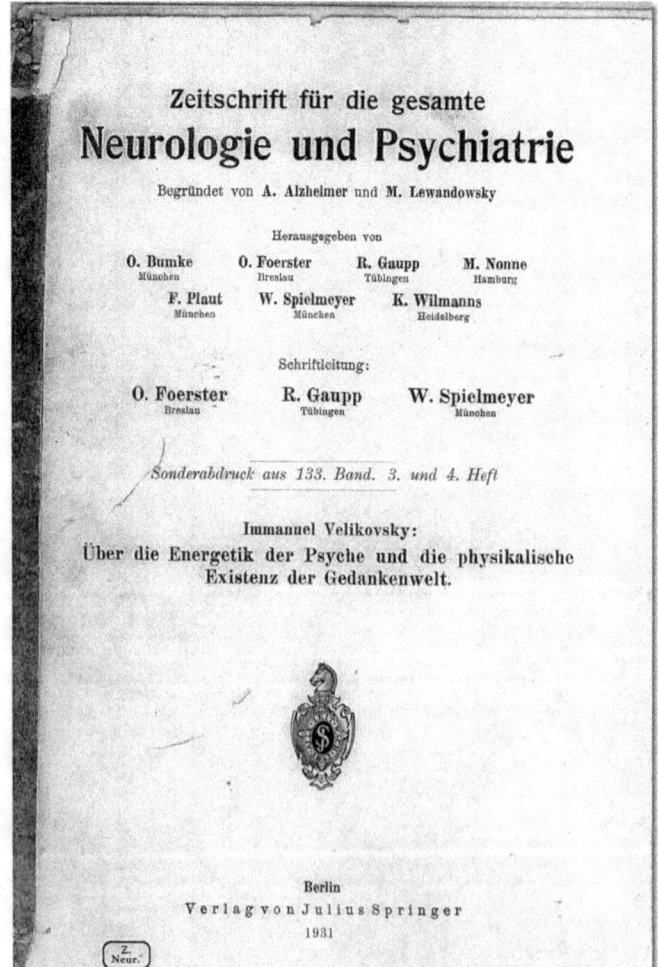

At the time of his mother's death, my father's focus shifted from general medical practice to a specialty in psychiatry. He was especially interested in the areas of the collective unconscious mind and the physical aspect of mental process.

In 1930 – 1931 Aba traveled to Zurich and Geneva to study neurology at the Monakow Brain Institute under Eugene Minkowsky. He, however, disagreed with Minkowsky's basic tenet that there is no connection between physical and psychic phenomena. Nevertheless, it was through his association with Minkowsky that my father met Eugene Bleuler, who was a pioneer in the area of schizophrenia.

Also in 1930 my father became acquainted with Carl Jung, a very tall man, who had a Great Dane by his side. During one conversation, Jung became angered when my father referred to him as Freud's disciple, since Jung had broken theoretical ties with Freud fifteen years earlier. Jung referred my father to a female analyst, but my father discontinued meetings with her shortly thereafter.

Letter 1 (transl.)

Dr. C. G. Jung   Küsnacht-Zürich
                                                        Seestreet 228, 28.3.30

Dear Mr. Dr.,

Unfortunately I will not be in Zurich in mid April: I will only begin again in early May. Then I can inform you about the possibility of an internship.

                    With high regards,
                    Yours sincerely,
                    C. G. Jung

In 1931 Aba authored a paper about the relationship between anger and thought, *On the Physical Existence of the World of Thought*. Just as in his later work his ideas were way ahead of his time. He believed that thoughts or ideas have physical existence of energy, which would explain the phenomena of telepathy. Also, in this paper, my father predicted that epileptics would exhibit different brain waves than normal people did, which he demonstrated in a test conducted on an epileptic boy in Tel-Aviv. Using a modified cardiograph, he recorded an early curve of petit mal. From this theoretical perspective my father advanced other hypotheses. He predicted that sight could be restored to the blind and hearing to the deaf if intact nerve centers in the brain could be artificially stimulated by impulses. Thirty years later, other scientists echoed my father's ideas.

Prof. EUGEN BLEULER
ZOLLIKON bei ZÜRICH

Tel. Zürich Limmat 96.69

ZOLLIKON, 23 VII 30
Zollikerstrasse 98

Herrn Dr. Im. Velikowsky   Poste rest. Genf

Sehr geehrter Herr College!
　Gegen Ihren Vorschlag zur Aenderung des ersten Satzes hätte ich einzuwenden, dass man im Deutschen nicht gern mit "Ich" anfängt. Aber wenn die erste Fassung aus einem Grunde, den ich nicht errate, Ihnen nicht gefällt, so kann ich ja diese Eitelkeitsstellung stehen lassen. Dafür aber möchte ich bitten, den Ausdruck "occult" nicht durch "übernormal" zu übersetzen. Ich halte den letzteren Begriff für falsch.
　Die zweite vorgeschlagene Aenderung können Sie in der Form: "wenn ich auch nicht mit allen Einzelheiten einverstanden bin", hineinsetzen. Ich finde keinen grossen Unterschied gegenüber meiner Fassung.

　　　　Mit collegialen Grüssen ergebenst

Letter 2

DR. MED. IMMANUEL VELIKOVSKY

Javne str., 31
Tel-Aviv, Palestine

Herrn Prof. Dr. E. Bleuler,

Sehr geehrter Herr Professor,

mit gleicher Post sende ich Ihnen meine Arbeit die in der Z. f. g. N. u. Ps. erschienen ist. Ich habe das Bedürfnis Ihnen noch einmal zu danken! Sie haben mich, Unbekannten, an die Hand genommen, und in die Wissenschaftliche Welt eingeführt.

　　　Ihr ganz ergebener

Falls es Ihnen genehm erwünscht ist dass die Arbeit an bestimmte Stellen geschickt wird, so bitte ich um Ihrer Mitteilung der Adressen.

Letter 3

Telepathy was considered controversial in the field of psychiatry. However, my father's paper was published in the leading neurological journal, *Zeitschrift für die gesamte Neurologie und Psychiatrie* (133, 1931). The preface was written by Eugene Bleuler, who wrote that Aba's theory "is not only stimulating but it may also help science overcome its unworthy shyness to probe into a new strange field." As though Bleuler had known what my father's future work would entail.

Letter 2 (transl.)

Mr. Dr. Velikovsky   23.7.30

Dear Colleague!
In response to your recommendation to change the first sentence, I would interject that in German one does not like to begin with "I". But if the first draft is not to your liking for a reason that I cannot guess, then I could accept this word order of vanity. But in exchange I would request not to translate the term "occult" with "übernormal" ("supernatural"). This term I deem incorrect.

The second recommended change could be in this form: "wenn ich auch nicht mit allen Einzelheiten einverstanden bin." ("even though I do not agree with all the details"). I find no great difference from my version.

With colleague-greetings respectfully
Bleuler

Letter 3 transl. (a draft written by Velikovsky)

Mr. Prof. Dr. E. Bleuler,

Under separate cover I send you an offprint of my paper that appeared in Z./g. N & Ps. I have the need to thank you once again: You took me – an unknown – by the hand and introduced me to the scientific world.

Yours very faithfully

In case you find that the paper ought to be sent to specific places, then I request your communicating the address to me.

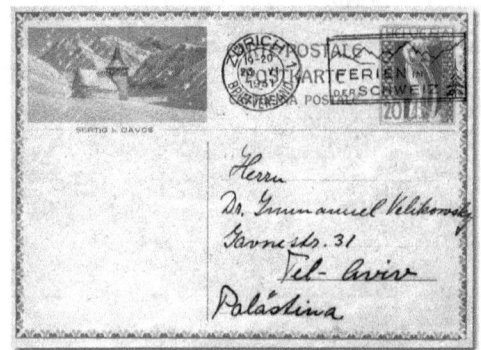

```
                                    Zollikon, 22 VI 31
Herrn Dr. Immanuel Velikowsky
       Javne Str. 31 Tel-Aviv, Palästina.

Sehr geehrter Herr College!
Verbindlichen Dank für die Uebermittlung des Sepa-
ratums. Ihrer freundlichen Offerte gemäss erlaube ich
mir, Sie um zwei weitere Exemplare zu bitten, die
ich Interessenten zuwenden möchte.
       Mit collegialen Grüssen ergebenst
```

Postcard 1

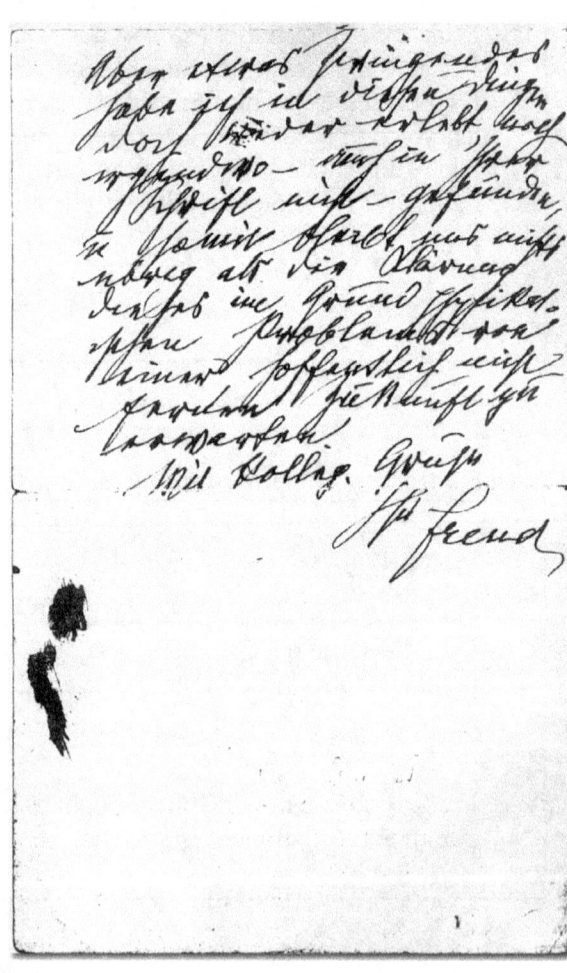

Letter 4

Aba and Psychoanalysis

Postcard 1 (transl.)

> Zollikon, 22. VI. 31
>
> Mr. Dr. Immanuel Velikowsky
> Javne Str. 31 Tel-Aviv, Palestine
>
> Very honored colleague!
> My best thanks for sending me the offprint. According to your friendly offer I take the liberty of asking you for 2 more copies, which I would like to give to interested persons.
>
> With colleague-greetings sincerely,
> Bleuler

Aba sent a copy of his paper to Sigmund Freud who responded by saying that he also had independently arrived at the same conclusions, a practice shockingly familiar of psychoanalysts taking credit for work of their disciples (Letter 4):

Letter 4

> Prof. Dr. Freud 24.6.1931
> Wien, IX, Berggasse 19
>
> Geehrter Herr Kollege,
>
> ich kann mich zum Inhalt ihres Aufsatzes (Energetik der Psyche) ganz übereinstimmend mit Bleuler äussern. Auch ich habe mir über den Gegenstand selbständig Meinungen gebildet, die den Ihren sehr nahe kommen, sich in manchen Stücken gradezu mit ihnen decken. Gegen eine energetische Auffassung der Denkprozesse hat grade der Analytiker an wenigsten einzuwenden. Eigene Erfahrungen haben mir die Vermutung nahe gelegt, dass die Telepathie der reale Kern der angeblichen parapsycholog. Phänomene ist und vielleicht der einzige.
>
> Mit kolleg. Grusse
> Ihr Freud

Letter 4 (transl.)

> Prof. Dr. Freud 24.6.1931
> Vienna, IX, Berggasse 19
>
> Honored Colleague,
>
> In response to the content of your paper, I find myself in total agreement with Bleuler. I too, have come to some independent conclusions concerning the subject, that approach yours very closely, even correspond to yours exactly in some parts. It is precisely the analyst who rejects an energetic understanding of thought processes least. Personal experience had led me to suspect that telepathy is the true core of so-called parapsychological phenomena and perhaps the only one.
>
> With colleague-greetings,
> Your Freud

Aba and Stekel

Ima and Stekel

PSYCHOTHERAPEUTISCHE PRAXIS

VIERTELJAHRESSCHRIFT
FÜR PRAKTISCHE ÄRZTLICHE PSYCHOTHERAPIE

HERAUSGEGEBEN VON
DR. W. STEKEL, WIEN / PROF. DR. A. KRONFELD, BERLIN
FÜR DIE SCHWEIZ:
DOZ. DR. O.-L. FOREL, PRANGINS / DOZ. DR. W. MORGEN-
THALER, BERN / PROF. DR. J. E. STAEHELIN, BASEL
FÜR DIE NORDISCHEN LÄNDER:
DR. POUL BJERRE, STOCKHOLM / DR. OLUF BRÜEL,
KOPENHAGEN / DR. HELGI TÓMASSON, REYKJAVIK
SCHRIFTLEITUNG: DR. ERNST BIEN, WIEN

| BAND 2 | MÄRZ 1935 | HEFT 1 |

INHALTSVERZEICHNIS

ORIGINALIA
Prof. Dr. A. Kronfeld: Zur Psychologie des Süchtig-Seins . . . 1
Dr. J. Velikovsky: Psychische Anaphylaxie 10
Dr. H. Rehder: Eine einfühlend geordnete Behandlung
    hysterischer Leiden 17

MITTEILUNGEN
Doz. Dr. A. Herzberg: Kasuistik der Psychotherapie
    bei älteren Leuten 24
Dr. W. Stekel: Ein Traum eines Arztes 32

VARIA
Dr. O. Brüel: Sexualsymbolik im volkstümlich-religiö-
    sen Kult 36
Prof. Dr. G. B. Gruber: Praktische Psychologie . . . 37
Dr. E. Bien: Die Behandlungsmethoden der Impotenz . 38
Prof. Dr. J. Kollarits: Tuberkulose, Charakter, Hand-
    schrift 40
Einige Hippokratische Lehrsätze 42
REFERATE 43
PSYCHOTHERAPEUTISCHER BRIEFKASTEN . . . 63

VERLAG DER PSYCHOTHERAPEUTISCHEN PRAXIS (WEIDMANN & CO.)
WIEN   LEIPZIG   BERN

SONDERABDRUCK

A year later, in 1932, Freud published some of these same ideas in his book *New Introductory Lectures*.

Twenty-five years later, my father discussed this incident in a paper titled, »Very Similar, Almost Identical«, in *Psychoanalysis and the Future* (New York 1957), p.14-17, 152-53.

Aba returned to Palestine in 1931, and we moved from the Carmel to Tel-Aviv, where he expanded his medical practice and served as chairman of the local Psychological Society. In 1933 my parents returned to Europe to be trained in psychoanalysis under Wilhelm Stekel who himself had been trained by Freud. My mother's one appointment with Stekel discouraged her from returning. Stekel, laughing at something she had said, upset her. My father met with Stekel for about three months.

The men in the Vienna psychoanalytic community were a significant influence in my father's life.

My father visited Freud on his birthday when he had already suffered from cancer of the jaw. He left his guests and sat on the porch with my father for about an hour. Freud, my father said, was a slow thinker. He was courteous to my father, both in person and in letters. Stekel, on the other hand, was a quick thinker. Stekel, as my father told it, wrote directly into the typewriter, and never read the manuscript before submitting it to the publisher.

Many years later, in 1978 speaking to my class at the Center for Modern Psychoanalytic Studies in New York, my father told an anecdote of Freud saying he would not invite Stekel to his home for fear the silverware would disappear, undoubtedly alluding to Freud's feelings that Stekel had stolen ideas from him. Reading Stekel, however, repeated credit is given to Freud.

My father presented Stekel with his own reinterpretation of Stekel's dream analysis in his book *Neurosis and Psychosis*. Stekel admitted that my father's view was the correct one. Several sessions later, Stekel told my father that he needn't continue seeing him, that he was a master and could train himself. In the lecture my father gave to my psychoanalytic class at the Center for Modern Psychoanalytic Studies in New York he said that that was to be his emotional undoing years later. Stekel published my father's paper, »Psychische Anaphylaxie«, a case study where my father linked a patient's asthmatic condition with a childhood trauma of near-drowning.

The meetings in Vienna consisted of training groups in analysts' homes, including Alfred Adler and August Aichhorn. Aba attended meetings of the International Psychoanalytical Society. At one of these meetings, he and Paul Federn were the only two to defend Freud's book on telepathy. Federn was a colleague of Freud's and had taken on his patients after Freud developed cancer. Federn and my father thereafter began a lifelong friendship.

IMMANUEL VELIKOVSKY

KANN EINE NEU ERLERNTE
SPRACHE ZUR SPRACHE DES
UNBEWUSSTEN WERDEN?

· SONDERABDRUCK AUS
IMAGO
ZEITSCHRIFT FÜR PSYCHOANALYTISCHE PSYCHOLOGIE
IHRE GRENZGEBIETE UND ANWENDUNGEN
BAND XX (1934) HEFT 2

[Reprinted from THE PSYCHOANALYTIC REVIEW, Vol. XXI, No. 3, July, 1934.]

CAN A NEWLY ACQUIRED LANGUAGE BECOME THE SPEECH OF THE UNCONSCIOUS?
WORD-PLAYS IN THE DREAMS OF HEBREW-THINKING PERSONS

By Dr. IMMANUEL VELIKOVSKY

TEL-AVIV, PALESTINE

The question of the identity of dream symbolism in languages of various origin is of far-reaching importance; its solution can cast light on many fields of psychology (origin of languages, ideation, collective unconscious, hereditary transmission of the mneme, etc.).

Since I often conduct my analyses in Hebrew, I believe that I may be able to contribute to the solution of the problem through a comparison of the symbolism in the Semitic and Aryan languages, which are further removed from one another than the various branches of the Aryan. As preliminary studies, I have decided to publish "Psychoanalytic Precursors in the Art of Dream Interpretation of the Ancient Hebrews from the Tractate Brachoth" (*Die Psychoanalytische Bewegung*, Vol. V, No. 1) and the present paper.

In this paper I want to show the existence of unconscious thinking in the Hebrew (a newly revived language, no hereditary transmission of the mneme) and, through examples, to present some idea of the richness of word-plays found in the dreams of individuals thinking in Hebrew, and at the same time to attempt an explanation for this frequent appearance of word-plays in Hebrew.

One idea can replace a second as the result of a similarity in form, qualities, functions, the manner of originating, or of a similarity in the sound of the word. The symbol (which arises through similarity in form, quality, function or origin) is not or not necessarily bound to any particular turn of speech. The greater the similarity of the symbol to the word-sound the less its effectiveness. The opposite is true of the word-play; it may be a success if there be only a similarity in the sound of the words and otherwise none whatsoever.

Since word-plays are intimately and essentially bound up with a particular language, they are different in every language.

The correct interpretation of symbols in a language which in this field has hardly been at all exploited is possible only after much

[329]

XIᵉ CONGRÈS INTERNATIONAL DE PSYCHOLOGIE

PARIS, 25-31 JUILLET 1937

IMMANUEL VELIKOVSKY
(Tel-Aviv)

Les Origines psychologiques
de la haine des Nations

Aba's Psychoanalytic couch in
Tel-Aviv

Freud published several of my father's papers. One, »On the Dream Interpretation in the Traktate Brachot of the Talmud« appeared in *Psychoanalytische Bewegung* V. 1 (1933), pp. 3-6. In the paper my father demonstrated that the ancient Hebrew practice of dream interpretation was similar to the 20th century psychoanalytic method. Before leaving Vienna, he also published a paper on the psychological function of the secretion of tears in relieving depression.

At this time, Aba continued experimenting with electroencephalography, while publishing several more papers on psychoanalysis, including a paper »Can a Newly Acquired Language Become the Speech of the Unconscious? Word Plays in the Dreams of Hebrew-Thinking Persons« about a dream and an entire neurosis based on a play on words. It was published in the *Psychoanalytic Review* 21 (July 1934), p. 329. While attending the International Congress of Psychologists in Paris in 1935, my father presented his paper, »The Psychological Origins of the Hatred among Nations«.

In Palestine, my father worked hard and long in his psychoanalytic practice. I remember an anorexic girl with long braids, who – having seen many doctors in Europe – began to gain weight under my father's care. She drank tea with us.

Anorexic Patient
before Treatment

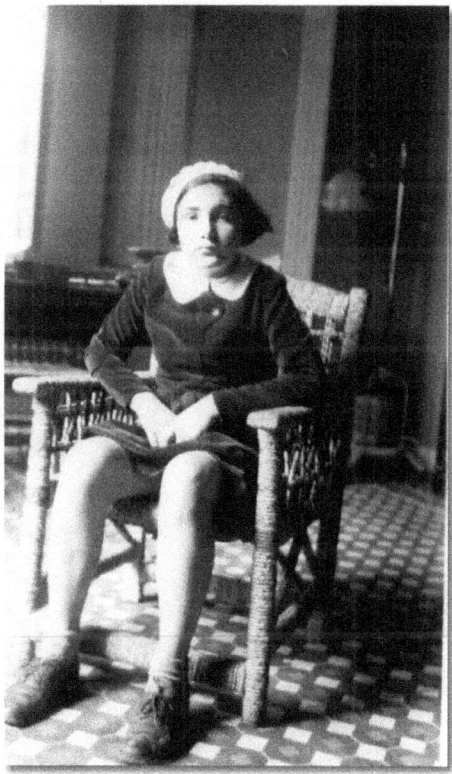

Anorexic Patient
after Treatment

Mittwoch, 21.?

Mein liebster Schwedi,

heute hatte ich ein Feiertag, da ich mit einmal 3 Briefe bekommen habe: von dir, von Sabeli, und von Drele. Gestern hatte ich auch ein Brief von dir mit Bildern. Geipele wo sie lacht und Lala so sie verträumt schaut – sind sehr gut. Auch Drele mit Kindern sind gut. Ich freue mich dass hela wieder bei Kindern ist.

Was soll ich dir von mir berichten? Eines habe ich fest entschlossen: auf keiner Sache in voraus zu verharren. Und so denke ich diese Tage, morgen oder übermorgen nach Genf (Geneve) fahren. Dort habe ich Baudoin, Hornoy und meinen Onkel Michall (?)aufgesucht. Der Onkel ist allerdings nur einige Jahre (ich denke 5) älter als ich. Bei Jung bleibe ich nicht. Schliesslich er hat keine eigene Gedanken. Er konnte die Mehrheit auf dem psychoan. Kongress (vor dem Kriege) bekommen, aber doch er ist geworden ohne Bewegung, da er keine eigene Gedanken hatte. Er musste sie nur schnell schaffen; so ist kann ich es auch. Mache bitte Du keine Sorge deswegen, es ist gerade ein Zeichen, dass ich selbständig denken kann, und nicht in verba magistri (Worte des Lehrers) schwöre. Dazu noch Zeit und Geld vergeuden bei seiner Ausdeutsch woll- te ich auch nicht: in dieser Reise muss ich alles

2

suchen um selbst zum Ausdruck kommen. Ich hatte noch einige mir scheint interessante Gedanken für meine Arbeit (Psych. Energie), und momentan bin ich in Glauben, dass da nun was sein; jedenfalls die Gedanken von Ostwald (und anderer) habe ich selbständig wiedererfunden, und es muss mich stärken in meinem Glauben. Dann Bleuler ist nur jetzt in Ferien und kommt in Drei Wochen wieder; ich hätte ihn nachgefahren in seine Ferien, aber es ist gerade in d. dritten Winkel des Dreiecks Schweiz (Zürich – Lugano – Genf). Dabei muss ich doch etwas klarer die Sache formulieren. In Genf habe ich Hornoy, er ist sehr bedeutend und klug.

Ausserdem will ich sehen ob nicht eine Möglichkeit besteht bei Liga der Nationen zu arbeiten. Ich habe gelesen, dass ein Professor aus Paris, der gerade vor kurzem ein Artikel über Judenfrage veröffentlicht hat, vor eingeladen zur Leitung eines planierten Institut für internat. Wissensch. Zusammenarbeit. Meine Phantasie sieht mich als einem (bezahlt.) Mitarbeiter. Allerdings nur die Phantasie. Möglich mein Onkel wird können mir was raten.

Mit dir so heute war ich zweites Mal bei Frau Einsalem. Sie ist gefahren um drei Kinder zu Andern, um ev. selbständig später in Palästina zu sein. Ihr Mann sollte nun früher da

My father proceeded with his pursuit of psychoanalytic training in Europe, while my mother remained in Palestine taking care of my sister and me. During this time my father wrote many letters to my mother.

Letter 5 (transl.)

My dearest Schewik   Wednesday, 21. V.

Today I had a joy when I received 3 letters at once: from you, from Sabele, and from Dorele. Yesterday I also had a letter from you with pictures. Geigele (Shulamit) laughing and Lala (Ruth) looking dreamy. They are very good. Also Dorele with children was nice. I'm happy that Hela is with the children again.

What shall I tell you about myself? I have determined one thing: never count on one thing in advance. And so I plan today or thereafter to go to Geneva. There I have Baudouin, Flornoy, and my uncle Michael Grodensky. However, the uncle is only a few years (5 I think) older than I am. I won't stay with Jung. After all, he does not have his own ideas. He was able to receive the majority during the psychoanalytic congress (before the war) but he has become dynamic-less, since he does not have his own thoughts. He just had to create them quickly – I can do that too. Please don't worry about that: it is precisely a sign that I can think independently and do not swear by the words of the teacher. Then, to waste time and money with his assistants is not what I wanted to do: during this trip I must do everything to express myself. I had some more interesting ideas concerning my work (Psych. Energy) it seems to me. At the moment, I believe there is something there; in any case I have reinvented the ideas of Ostwald (and others) independently and that must strengthen me in my belief. Bleuler is on holiday now and will return in two weeks. I would have followed him in his vacation, but he is just in the third corner of the Swiss Triangle (Zurich-Lugano-Geneva). Besides, I have to formulate the thing a little more clearly. I have Flornoy in Geneva – he is very important and smart.

Also I want to see if there is no opportunity to work at the League of Nations. I read that a professor from Paris, who had recently published an article on the Jewish question, was invited to head a newly planned Institute for International Scientific Cooperation. My imagination sees me as a (paid) staff member. That is, however, just my imagination.

Today I visited Mrs. Eusalem a second time. She went to study diet-cooking, in order to become independent later in Palestine. Her husband should now ...

(The rest of this letter is missing)

Montag.

Mein guter, guter, lieber, liebster Schewik.
Und so bin ich wieder in Luzern. Die kleine Schäry
habe ich noch nicht gesehen. Hoffentlich wird sie
gut angenommen. Beschreibe mir, wie hast Du
Selma getroffen. Ich wünsche dass sie soll Dich
mit Freundlichkeit anstecken, weil ich mache mir
ein bisschen Vorwürfe, dass ich nicht auf der Höhe, besonders erste Woche war. *am ersten Hübner*
Ich bestehe darauf dass Du sollst täglich Fleisch
essen, möglich auch Leber (es ist gut bei Blutarmut), auch
viel Spinat, Früchten, Milch. Ich bitte Dich sehr
deswegen. Hast Du schon Eisen Carson gekauft?
Jetzt habe ich an Michael geschrieben, ins Jeu
geschrieben (dass ich komme), und der Tante Jenny
nach Spa; ich habe ihr auch geschrieben,
dass Du müde bist und deswegen wirst viel
ausruhen und weniger umreisen.
Grüsse bitte Selma. In warmer, warmer Liebe
Dein Emanuel
Sei mutig, lustig, gesund, frisch, vergnügt.
Was für dummes Sprach ich auf dem
Schiff, ich schäme mich. Sei mir gut, liebster
Schewik.
(In Baden-Baden ist Neumann; kannst Dich zu ihm
immer wenden.)

Letter 6

Italy   Paris

Aba and Psychoanalysis

Letter 6 (transl.)

> My good, good, dear, dearest Schewik,   Monday
>
> And there I am again in Zurich. I haven't seen the little Schärf (?) yet. Hopefully you have arrived well. Describe me, how you have met Selma. I wish she should infect you with happiness; because I am reproaching somewhat to myself that I didn't feel up to the mark, especially the first week.
> I insist that you eat meat daily, chicken would be best, if possible also liver (this is good with anemia, also a lot of spinach, fruit, milk. I implore you a lot about this. Have you bought Iron Elasson (?) yet?
> I have now written to Michael Grodensky in Geneva (that I'm coming) and to Aunt Jenny at Spa; I have also written that you are tired and therefore are going to rest much and travel little.
> Give my regards to little Selma. With warm, warm love
>   Your Emanuel
>
> Be brave, cheerful, healthy, fresh, jolly. What kind of stupidities I said on the ship, I am ashamed. Be good to me, dearest Schewik.
> (At Baden-Baden there is Neumann; you can always address yourself to him.)

In 1937, Shulamit and I joined our parents on one of their trips to Europe, visiting France, Italy and Switzerland.

In New York, after getting permission to practice, my father saw patients in an office adjacent to my mother's sculpture studio at 558 West 113th Street.

In Princeton my father continued to see a few patients who came from New York. When I became a psychoanalyst I was frequently amazed to hear how progressive my father's approach to psychoanalysis was.

Stekel wrote to Chaim Weizmann, then president of Israel:

> The university in Jerusalem must now become the "guardian place" of the Jewish science (Psychoanalysis). You have in Tel-Aviv one of the most highly gifted of psychotherapists, Dr. Velikofsky [sic], with whom it would be worthy for all scholars to unite, and in Jerusalem to create a center for Germany's prohibited branch of psychoanalysis.
>   Wilhelm Stekel to Chaim Weizmann in Ronald W. Clark's *Freud: The Man and The Cause*, 1980, New York; Cape/Weidenfeld&Nicholson, p. 493.

Postcard 2

> Sehr geehrter Herr Doktor,
> ich sende Ihnen anbei eine Arbeit für Imago ("Psychvan. Ahnungen"). Ich werde Ihnen verpflichtet sein, wenn Sie mich bald benachrichtigen würden, ob und wann die Arbeit erscheinen wird. Erhält man Sonderdrucke und wieviel?
> Ich benutze die Gelegenheit und sende für d. Referatenteil ein Separatum (mit Vorwort v. E. Bleuler)
> Mit vorzüglicher Hochachtung
> Ihr ergebener
> Dr. Im. Velikovsky

Letter 7

On the following pages you find some examples of my father's psychoanalytic correspondence.

Postcard 2 (transl.)

> With colleage-greetings
>
> Prof. E. Bleuer
>
> Zollikon near Zurich
> Zollikerstr. 98
> 18. VII. 30

Letter 7 (transl.)

> 30.8.32
>
> Very Honored Doctor,
>
> I am sending you with this letter a paper for Imago ("Psychoan. prescience …"). I'll be very much obliged to you, if you inform me soon, whether and when the paper will be published. Will I get offprints and how many?
> I am using this opportunity and am sending you an offpint (with preface by E. Bleuler) for your review section.
>
> Respectfully
> Dr. Im. Velikovsky

IMAGO
ZEITSCHRIFT FÜR PSYCHOANALYTISCHE PSYCHOLOGIE, IHRE GRENZGEBIETE UND ANWENDUNGEN

REDAKTION:
DR. ERNST KRIS
UND
DR. ROBERT WÄLDER

HERAUSGEGEBEN VON
SIGM. FREUD

ALLE ZUSCHRIFTEN AN:
INTERNATIONALER
PSYCHOANALYTISCHER VERLAG
WIEN I. BÖRSEGASSE 11

Dr.K/R.

Wien, 25. Oktober 1932.

Herrn

Dr. Immanuel V e l i k o v s k y

Tel Aviv

Sehr geehrter Herr Doktor!

Dr. Fenichel hat uns den Sonderabdruck Ihrer Arbeit über Energetik etc. und Ihr Manuskript über "Psychoanalytische Ahnungen in der Traumdeutung der alten Hebräer nach dem Traktat Brachoth" zugesandt. Ein Referat Ihrer Arbeit wird ehestens erscheinen und Ihr interessanter Aufsatz entweder in der "Imago" oder in der "Psychoanalytischen Bewegung" zum Abdruck gelangen.

Wir danken Ihnen nochmals für Ihre beiden Zusendungen, hoffen auch weiterhin mit Ihnen im Kontakt bleiben zu können und sind

mit den besten Empfehlungen

Ihre ergebenen

Kris   R. Wälder

Letter 8

P. B. 194, Tel Aviv
4. XI. 32.

Herrn Dr. E. Kris und Herrn Dr. R. Wälder,
Redaktion "Imago".

Sehr geehrte Herren Kollegen

ich erhielt Ihr frdl. Schreiben v. 25. X. Zu der Referierung meiner Arbeit über Energetik: diese Arbeit war von Prof. Freud in einem Brief an mich besprochen; er äusserte sich, dass er mit Bleuler (in seiner Vorrede) ganz übereinstimmt, und dass mehrere Gedanken der Arbeit auch bei ihm entstanden sind. Eine eingehende Kritik aus der Hand von Prof. Freud, wenn er dazu einwilligt, würde einmal seiner Stellung zu diesem wichtigen Gebiet der Psychologie Ausdruck geben.

Von der für Imago bestimmten Arbeit (Psych. Ahnungen der alt. Hebr.), die Sie freundl. zum Abdruck aufgenommen haben würde ich wenn möglich gerne 1) Korrektur Blätter einmal lesen (in 12-14 Tage in Wien zurückerhaltbar) 2) eine Anzahl von Separatis, ev. auf meine Kosten erhalten. Von verbleibe Ihr sehr ergebener

Dr. Im. Velikovsky

Letter 9

Letter 8 (transl.)

Vienna, 25. Oct. 1932

Dear Dr. Velikovsky,

Dr. Fenichel has sent us a reprint of your article on Energetics etc. as well as your manuscript on "Psychoanalytical Prescience in the Art of Dream Analysis of the Old Hebrews According to the Treatise of Brachoth." A report of your work will appear shortly and your interesting paper will appear in "Imago" or in "Psychoanalytische Bewegung (Psychoanalytic Movement)."
We thank you again for sending us these materials and hope to remain in contact with you further.

> With best wishes,
> Respectfully
>
> Kris   Wälder

Letter 9 (transl.)

P.B. 194 Tel-Aviv
4. XI. 32

Dr. E. Kris and Dr. Wälder
Editorial office "Imago"

Very honorable colleagues,   -
I received your friendly letter dated 25.10 on the peer review of my paper on Energetics: this paper was reviewed by Prof. Freud in a letter to me. He commented that he was in full agreement with Bleuler (in his preface) and that several of the ideas in the paper have also come to him. An extensive critique by Prof. Freud, if he is willing, would be an expression of his views on this important subject of psychology. Of the paper destined for Imago (Psychoan. Prescience of the Old Hebrews) which you have so kindly selected for publishing, I would like to have – if possible – 1. corrections for review (to be returned to Vienna in 12-14 days) and 2. a number of reprints, if necessary on my account.

> I remain your faithfully,
> Dr. Im. Velikovsky

DIE PSYCHOANALYTISCHE PRAXIS

VIERTELJAHRESSCHRIFT FÜR DIE AKTIVE METHODE DER PSYCHOANALYSE

VERLAG: S. HIRZEL, LEIPZIG — SCHRIFTLEITUNG: Dr. WILHELM STEKEL, WIEN-SALMANNSDORF

Wien 23/I 1934

Lieber Freund:

Ich freue mich ganz ausserordentlich, dass Ihre Frau jetzt vor die Oeffentlichkeit tritt und Ihr grosses Können zeigen und Menschen begeistern kann.

Die Ueberflutung von Palestina mit Psychotherapeuten war vorauszusehen.

Die Deutsche Vereinigung der Psychotherapeuten hat ein Jung die Führung übergeben, der im Vorwort betont, dass man jüdische Psychologie von der christlichen trennen müsse....

In einem zweiten Vorwort wird verlangt, dass jeder Psychotherapeut das Buch von Hitler "Mein Kampf" genau studieren müsse....

Unsere Zeitschrift ist in unseren eigenen Verlag übergegangen und wird also eine ausgesprochen jüdische Zeitschrift sein. Sie heisst jetzt:

    Psychotherapeutische Praxis.

Wir wollen sie auf eine breitere Basis stellen.

Das erste Heft wird die Organneurosen behandeln.

Ich möchte, dass Sie schon im ersten Hefte vertreten sind und ersuche um eine praktische Arbeit.

Theoretische Arbeiten passen nicht in unseren Rahmen.

Wir wollen die praktischen Aerzte erobern.

Wir haben unser Ambulatorium eröffnet und wir ziehen die Aerzte heran. Das Material ist sehr gross. Wir halten schon bei Nummer 210.

Ich lese jede Woche für Aerzte, haben ausserdem noch einen privaten Kurs für 5 Aerzte, die aus dem Ambulatorium Fälle unter meiner Führung analysieren.

Von Gutheil und Wengraf erscheinen neue Bücher.

Von mir: Die Erziehung der Eltern.

Auch die anderen Herren halten teils populäre teils wissenschaftliche Vorträge.

Materiell geht es sehr schlecht.

Auch die politische Situation ist sehr unsicher und ich verstehe, dass jeder Jude den Wunsch hat, nur unter Juden zu leben.

Har Even tut mir leid. Politisch Lied ein garstig Lied.

Der Zug zum Extremen müsste psychologisch erklärt werden.

Bitte mir recht bald eine praktische Arbeit (eventuell Kasuistik-ihre Erfolge mit oder ohne Analyse) zu senden.

Herzliche Grüsse von Haus zu Haus

Ihr getreuer

Wilhelm Stekel

Letter 10 (transl.)

Vienna 23. I. 1934

Dear Friend;

I am extraordinarily glad that your wife now steps out to the public and shows her great capacity, as well as making people enthusiastic.

It was to be expected that Palestine would be run over with psychotherapists.

The German Association of psychotherapists has handed over guidance to Jung who emphasizes in the preface that one has to separate the Jewish psychology from the Christian one ...

In a second preface, it is required that every psychotherapist has to study Hitler's book "Mein Kampf" thoroughly ...

Our journal has been taken over into our own publishing house and therefore will be a downright Jewish Journal. It is now called Psychotherapeutic Practice.

We want to put it on a wider basis.

The first issue will be about organ neurosis.

I would like you to be represented in the first issue and beg you for a practical contribution.

Theoretical contributions do not fit our frame.

We want to capture the practical doctors (medical doctors who specialize in common practice).

We have opened our ambulatory practice and we are putting in the doctors.

The material is very large. We stopped at number 210.

I give lectures every week for doctors and we have besides also a private course for five doctors, who analyze cases from the ambulatory practice under my guidance.

There are new books appearing by Gutheil and Wengraf.

By myself: The Education of Parents.

Also, the other gentlemen are giving lectures in part popular ones and in part scholarly ones.

In material ways, I am doing very poorly.

Also, the political situation is very insecure and I understand that every Jew has the wish to live only among Jews.

Har Even makes me sorry. The Political Song is an ugly Song.

The trend to the extreme would have to be explained in psychological terms.

Please send me very soon a practical paper (perhaps successful cases with or without analysis).

Heartfelt greetings from house to house

Your faithful

Wilhelm Stekel

Wien 20/II 1934

Lieber Freund: Ich habe heute einen entsprechenden
Brief an W. abgeschickt und hoffe, wir werden einen
guten Erfolg haben.
Wir haben sehr unruhige Zeiten hier gehabt und hoffen
dass wir nun verschont sein werden. Ich habe momentan
sehr viel zu tun, aber meine Gesundheit lässt zu
wünschen übrig. Leichte Angina pectoris zu allen
anderen Zores....
Wir erwarten mit Spannung eine Arbeit für die Zeitschrift. Aber nur praktisch !Keine Theorien.
Die neue Zeitschrift kommt in einigen Wochen heraus.
Sonst nichts Neues. Grüssen Sie alle Freunde und
Ihre liebe Frau und seien Sie selbst herzlichst
gegrüsst von
  Ihrem
              Dr. Wilhelm Eitler

Dr Eitl !

Letter 11

IMAGO
ZEITSCHRIFT FÜR PSYCHOANALYTISCHE PSYCHOLOGIE.
IHRE GRENZGEBIETE UND ANWENDUNGEN

REDAKTION:	HERAUSGEGEBEN VON	ALLE ZUSCHRIFTEN AN:
DR. ERNST KRIS		INTERNATIONALER
UND	SIGM. FREUD	PSYCHOANALYTISCHER VERLAG
DR. ROBERT WAELDER		WIEN I, BÖRSEGASSE 11

Dr.W/H.   Wien, 31.März 1934.

Herrn

Dr.Med.Immanuel Velikovsky

                                    T e l - A v i v

Sehr geehrter Herr Doktor !

         In Beantwortung Ihres Schreibens vom 5.III. teilen
wir Ihnen mit,dass sich zu unserem Bedauern die Publikation Ihres
Aufsatzes etwas verzögert hat; wir haben seinerzeit den Umfang
des vorhandenen und angenommenen Materials ein wenig unterschätzt.
Insbesondere ist auch durch das Ableben des Präsidenten unserer
Ungarischen Vereinigung,Dr.Sandor Ferenczi, und durch die Not-
wendigkeit,für die Gedenkartikel für Ferenczi sofort Raum zu
schaffen,das ganze Publikationsprogramm ein wenig verschoben
worden. Wir werden Ihnen gerne die Korrekturen Ihres Aufsatzes
nach Tel-Aviv senden und haben uns auch Ihre Separatawünsche
vorgemerkt.

         Mit den besten Grüssen sind wir

                                    Ihre sehr ergebenen

                      Kris   R Waelder

Letter 12

Letter 11 (transl.)

Vienna 20. II. 1934

Dear Friend,

Today, I have sent a corresponding letter to W and hope we will be successful.

Our times are very unsettled here and we hope we will be spared. I am very busy now, but my health leaves something to be desired. A mild angina pectoris along with other worries ...

We await with eagerness a paper for the journal. But only practical. Nothing theoretical.

The new journal will appear in a few weeks. Apart from that no other news. Greet all friends and your dear wife, and let yourself be greeted heartily.

Your Dr. Wilhelm Stekel

In a rush!

Letter 12 (transl.)

Vienna, 31. March 1934

Very Honored Doctor,

In answer to your letter from the 5th of March, we would like to inform you that to our regret the publication of your paper will be somewhat delayed. Previously, we underestimated the amount of the present and accepted material a bit. In particular, through the passing away of the president of our Hungarian Society, Dr. Sandor Ferenczi, and the immediate need to create space for the articles in memory of him, the entire publication schedule has been moved back a bit. We will be pleased to send you the corrections of your paper to Tel-Aviv and have made a note concerning your wishes for offprints.

With best wishes we are sincerely,

Kris   R. Wälder

Wien 17/X 1934

Lieber Freund Velikovsky :
Ihre Arbeiten habe ich mit grossem Interesse gelesen, sie sind beide angenommen, da sie sehr schön und wertvoll sind. Ich vermisse in Ihrem Briefe Nachrichten, wie es Ihnen und Ihrer lieben Frau geht. Sie sind ja jetzt überfüllt mit Aerzten in Palestina, wodurch der Lebenskampf immer schwerer wird. Wir haben jetzt Ruhe in unserem Lande, aber die Praxis geht schwach und ich lebe hauptsächlich von Aerzten, die hierher kommen, um bei mir zu lernen. Ich kann mit Stolz behaupten, dass meine Schule immer mehr Anhänger gewinnt. Gesundheitlich geht es bei mir schwankend, aber im Grossen und Ganzen muss ich zufrieden sein.
Ich danke für die Uebersendung des Separatums, ich habe die Arbeit schon vorher gelesen und sie kommt ins das Archiv unseres Institutes.
Viele herzliche Grüsse von Haus zu Haus Ihr getreuer

Dr. Wilhelm Stekel

Letter 13

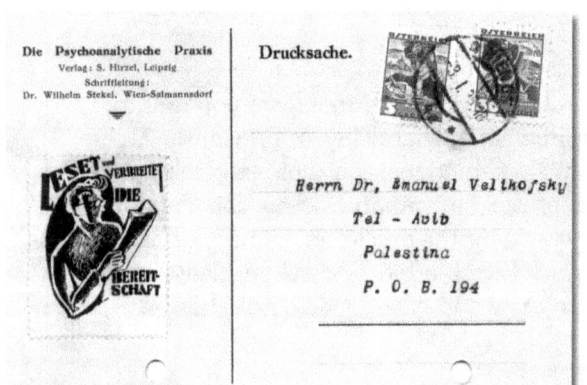

Wien 3/I 1935

Lieber Freund Velikofsky : Ich danke Ihnen herzlichst für Ihre grossmütige Spende. In diesen schweren Zeiten heisst das sehr viel...Hier ist es jetzt ruhig, wir athmen auf. Die Praxis geht schwach. Aber ich habe sehr gute Schüler und meine Lehre scheint sich auszubreiten- besonders in klinischen Kreisen. Ich habe jetzt zwei Schweizer Psychiater und einen Wiener. Das Ambulatorium ist sehr frequentiert und die Schüler arbeiten fleissig. Ich habe ein neues Buch beendet:Die Fortschritte der Traumdeutung. Es wird schon gedruckt und wird meiner Schule und jedem Analytiker sehr viel helfen. Die Erziehung der Eltern ist auch ein grosser Erfolg. Wann werden meine Briefe an eine Mutter in hebräischer Sprache erscheinen ? ... Sie sehen ich arbeite sehr fleissig. dabei wird heftig musiziert. Leider fehlt mir eine Künstlerin wie es Ihre liebe Frau ist. Mit vielen herzlichen Grüssen von Haus zu Haus Ihr getreuer

Dr. Wilhelm Stekel

Postcard 3

Letter 13 (transl.)

>   Vienna 17. Oct. 1934
>
> Dear Friend Velikovsky,
> I read your works with great interest; they have both been accepted, as they are very beautiful and valuable. I missed news in your letter about how you and your dear wife are doing. Palestine is now flooded with doctors which makes the struggle for survival more and more difficult. We now have peace in our country, but the practice is going slowly and I live mainly off doctors who come here in order to learn with me. I can proudly report, that my school is winning more and more followers.
> My health is waffling back and forth but all in all I have to be satisfied.
> Thank you for sending me the material under separate cover; I had already read the paper before and I will put it in the archive of our institute.
> Many heartfelt greetings from house to house. Your faithful
>   Dr. Wilhelm Stekel

Postcard 3 (transl.)

>   Vienna 3. I. 1935
>
> Dear Friend Velikovsky,
>
> I thank you dearly for your generous donation. In these hard times this means a lot ... It is quiet here now, we breathe more easily. My practice is going poorly. But, I have excellent students and my teaching seems to be spreading – especially in clinical circles. I now have two Swiss psychiatrists and one Viennese. The ambulatorium is heavily visited and the students work hard. I finished a new book: The Progress of Dream Analysis. It is already being printed and will help my school and any analyst quite a bit. The Education of Parents has also met with big success. When will my letters to a mother appear in Hebrew? ... You can see, I am working very hard. Also, there is lots of music-making. Unfortunately, I have no accomplished artist like your dear wife.
> With many heartfelt greetings from house to house,
> Your faithful,
>
>   Dr. Wilhelm Stekel

IMAGO
ZEITSCHRIFT FÜR PSYCHOANALYTISCHE PSYCHOLOGIE, IHRE GRENZGEBIETE UND ANWENDUNGEN

REDAKTION:	HERAUSGEGEBEN VON	ALLE ZUSCHRIFTEN AN:
DR. ERNST KRIS UND DR. ROBERT WÄLDER	SIGM. FREUD	INTERNATIONALER PSYCHOANALYTISCHER VERLAG

Dr.Kr./N.   Wien, den 28. Januar 1937

Herrn
Dr. Im. Velikovsky
P.O.B. 194
Tel-Aviv, Palestine.

Sehr geehrter Herr Doktor,

    Sie haben recht, uns zu mahnen, leider hat sich die Antwort so verzögert, weil bis vor kurzem unentschieden war, ob wir Ihre Arbeit nicht für den Almanach verwenden könnten. Wir denken jetzt doch daran, sie eher in der Imago zu bringen, müssen uns jedoch an mehreren Stellen Kürzungen vorbehalten. Sie wird aller Vermutung nach im Laufe des Jahres erscheinen. Wir werden Ihnen erst die Fahnen zur Einsicht vorlegen, da wir nicht annehmen, dass die Kürzungen Fragen betreffen werden, in denen Sie anderer Meinung sein können als wir.

    Mit der Bitte, die Verspätung unseres Schreibens zu entschuldigen, und mit besten Grüssen

sind wir Ihre ergebenen

Kris   R. Wälder

Letter 14

Letter 14 (transl.)

>Vienna, 28. January 1937
>
>Very Honored Doctor,
>
>You were right to admonish us. Unfortunately, our answer was delayed so much, because until recently, we were undecided whether we might not be able to use your paper for the Almanach. We are now tending toward including it in Imago, but must reserve the right to undertake some condensation in several places. According to our best estimate, it will appear in the course of the year. We will send you only the first printer's copy to be reviewed, as we do not assume that the condensation will address questions in which you could be of a different opinion than we.
>
>With the request, to excuse our delayed response and with best wishes,
>
>>We remain your faithfully,
>>
>>Kris   R Wälder

PSYCHOTHERAPEUTISCHE PRAXIS
VIERTELJAHRESSCHRIFT FÜR PRAKTISCHE ÄRZTLICHE PSYCHOTHERAPIE

HERAUSGEGEBEN VON PROF. DR. A. KRONFELD, BERLIN / DR. W. STEKEL, WIEN
FÜR DIE SCHWEIZ: DOZ. DR. O.-L. FOREL, PRANGINS / DOZ. DR. W. MORGENTHALER, BERN / PROF. DR. J. E. STAEHELIN, BASEL
FÜR DIE NORDISCHEN LÄNDER: DR. POUL BJERRE, STOCKHOLM / DR. OLUF BRÜEL, KOPENHAGEN / DR. HELGI TÓMASSON, REYKJAVIK
SCHRIFTLEITUNG DR. ERNST BIEN, WIEN

SCHRIFTLEITUNG

Wien, am 2.2.1937

~~IX. BERGGASSE 13~~
~~Telephon A-15-0-03~~
IX. SPITALGASSE 17

Herrn
Dr. Immanuel Velikovsky,
<u>Tel Aviv</u>

Lieber Herr Kollege !

Herzlichen Dank für Ihren Brief v.21.Jänner. Ich würde Ihre Arbeit über Zeichnungen neurotisch Kranker gern für die "Praxis annehmen; ich wüsste sonst keine Zeitschrift, die momentan hierfür in Betracht käme. Aber leider muss ich Ihnen mitteilen, dass grosse Gefahr besteht, dass unsere Zeitschrift im Eingehen begriffen ist, wenn nicht im letzten Augenblick Abhilfe kommt. Der Verlag Weidmann et Co. hat keine Lust und vielleicht kein Geld, die Zeitschrift weiter herauszugeben, so dass vom Jahrgang 1936 die Hefte 3 und 4 noch nicht erschienen sind. Es ist klar, dass wir alle grosses Interesse an der Erhaltung der Zeitschrift haben und deshalb wende ich mich an Sie mit der Bitte, Sie mögen als Abonnent der Zeitschrift an den Verlag Weidmann et Co. eine Mahnung bezgl. der zwei fehlenden Hefte richten, evtl. mit einer Klage drohen bezw. nach Erfolglosigkeit dieser Bemühung auch tatsächlich die Klage einreichen. Wir nehmen an, dass ein solcher Vorgang, von mehreren Abonnenten geübt, zum Erfolg führen könnte. <u>Ich bitte Sie selbstverständlich diese Anregung als vertraulich zu betrachten</u>.

Mit besten Grüssen
Ihr sehr ergebener

Bien

Letter 15

Wien 20/IV 1937

Lieber Freund Velikovsky:
Ich bin schon lange aus Brasilien zurück und werde mich sehr freuen, Sie zu sehen und Ihre Fortschritts der Erkenntnis kennen zu lernen. Habe schon Verschiedenes von Ihnen gelesen und freue mich, dass Sie so selbstständig Ihrer Wege gehen. Ich habe in Brasilien eine sehr schöne Zeit gehabt und nach Brasilien sehr viel zu tun, sowohl in der Praxis als auch als Lehrer. Ich habe jetzt viele neue Schüler. (Mit Bien habe ich aus verschiedenen Gründen gebrochen.) Habe einen sehr gut besuchten Kurs (22 Schüler), der mir viel Freude macht. Natürlich wird viel musiziert. Hätte gerne mehr von Ihnen gehört. Gehe nicht nach Paris, da ich Kongresse hasse und Paris um diese Zeit zu heiss ist. Dies in aller Eile mit vielen herzlichen Grüssen von Haus zu Haus Ihr getreuer

W. Stekel

Letter 16

Letter 15 (transl.)

Vienna, 2. 2. 1937

Dear Colleague!

Many thanks for your letter dated 21st of January. I would very much like to accept your paper on drawings by neurotically ill persons for the "Praxis"; I'd know no other appropriate journal at the moment. But unfortunately I have to tell you that there is the great danger that our journal will fold, if there isn't any help in the last minute. The publishing house Weidmann & Co. doesn't feel like continuing the journal, and maybe doesn't have the money for it either. So issues 3 and 4 of 1936 haven't been published yet. It is clear that all of us are very much interested in keeping the journal alive and therefore I am addressing you asking that you as a subscriber to the journal write a reminder about the two missing issues to the publishing house Weidmann & Co. Maybe you could threaten to file a lawsuit and even do so in case your efforts remain fruitless. We assume that such a procedure adopted by several subscribers, could be successful. Of course I must ask you to keep this proposal confident.

With best greetings
Your very faithful

Letter 16 (transl.)

Vienna 20. IV 1937

Dear friend Velikovsky,

I have long since returned from Brasil and will be happy to see you and to learn of your progress and insights. I have already read various things of yours and am pleased that you are taking such an independant course. I had a very pleasant time in Brasil and have much to do after Brasil, both in the practice and as a teacher. I have many new students now. (For various reasons I have broken contact with Bien). I have a well-visited course (22 students), that I thouroughly enjoy. Of course we make much music. Would like to hear more from you. Won't go to Paris as I despise conferences and as Paris is too hot at this time. This in great haste with heartfelt greetings from house to house. Yours faithfully,

W. Stekel

Aba and Psychoanalysis

Before going to the United States in 1939, my father had a nearly finished manuscript, titled *Introgenesis*, accepted for publication by Presses Universitaires of France, was working toward establishing a Jerusalem Academy, and had a busy psychoanalytic practice. But, despite his intellectual and Zionist achievements, at the age of 43, my father did not believe that he had achieved something significant as a scholar. Hence, at that time his focus shifted to a new scholarly endeavor, a book to be titled *Freud and His Heroes*. Part of this book would represent a reinterpretation of Freud's dreams. This he describes in *Stargazers and Gravediggers*:

> I found that his own dreams, sixteen in number, interspersed among numerous dreams of his patients in his classic *The Interpretation of Dreams*, spoke a language that was very clear but had meaning which Freud did not comprehend – or did not reveal to his readers. All the dreams dealt with the problems of his Jewish origin, the tragic fate of his people, his deliberations on leaving the ranks of the persecuted for the sake of unhampered advancement – or at least in order to free his children from the fate of underprivileged Jews in Christian and anti-Semitic Vienna. From this conflict, in which he struggled with himself he emerged victorious ... about the time when, unknown and obscure, he wrote his book on dreams.

The rest of the book was to challenge Freud's book, *Moses and Monotheism*. Freud believed that Akhnaton, not Moses, was the first monotheist. Moses, he claimed, was not Jewish, but Egyptian, and borrowed the monotheistic belief system from the Pharaoh Akhnaton. My father disagreed, saying that Akhnaton was actually a prototype of Oedipus.

In order to complete this project, which involved much research, my father needed access to libraries with extensive collections, which were absent in Tel-Aviv. Hence, the decision was made to travel to the United States and to stay there for eight months.

Publications and Drafts by Dr. Immanuel Velikovsky
Before his Departure to the U.S.

- Über die Energetik der Psyche und die physikalische Existenz der Gedankenwelt
  (On the energetics of the psyche and the physical being of thoughts)
- Perception and Conception
- Psychoanalytische Ahnungen in der Traumdeutungskunst der alten Hebräer nach dem Traktat Brachoth
  (Psychoanalytical prescience in the art of dream analysis of the old Hebrews according to the treatise of Brachoth)
- Eine Arbeitstheorie zum Verständnis der Melancholie und zu ihrer Behandlung
  (A working theory for understanding melancholy and for its treatment)
- Der Oedipus Komplex im Assoziationsversuch
  (The Oedipus Complex in association experiments)
- Can a newly acquired language become the speech of the Unconscious?
- Wortspiele in Träumen der Hebräisch Denkenden
  (Plays on words in dreams of people thinking in Hebrew)
- Über Spaltung der Persönlichkeit und die bewusste Verheimlichung in der Analyse. Margarethe-Neurose. Revision einer Analyse
  (On split personalities and its unconscious hiding during analysis. Revision of an analysis)
- Über verschiedene Intelligenzstufen in einer Person. Zum Problem der Abhängigkeit des Denkens von den Ausdrucksformen.
  (On various levels of intelligence in a single person. On the problem of the dependence of thought on the form of expression)
- Reflexe und Automatismen
  (Reflexes and automatisms)
- Psychogene Gehstörungen als Conversions-Merkmal bei den Analerotikern
  (Psychogenic walking-disturbances as a conversion-symptom among anal-erotics)
- Die Kreutzer Sonate von Tolstoi im Lichte der Lehre von der unbewussten homosexuellen Neigung
  (Tolstoy's Kreutzer-sonata in the light of the theory of subconscious homosexual tendencies)
- Analerotiker und ihre Lebensäusserung (Sprache, Handlung, Träume, Neurose)
  (Anal erotics and their living expression (speech, action, dreams, neurosis))
- Der Urtrieb. Die Similationsgewalt. Introgenese.
  (The primal drive. The Power of Similation. Introgenesis)

- Similation und Desimilation
  (Similation and Dissimilation)
- Über die vierfache Unsterblichkeit
  (On fourfold immortality)
- Kritik der Unzerstörbarkeit unseres Wesens von Schopenhauer
  (A critique of the indestructibility of our being by Schopenhauer)
- Der freie und der unfreie Wille
  (Free and un-free will)
- Der Fehler in dem Gesetze der Erhaltung der Energie
  (The error in the law of conservation of energy)
- Über die Entstehung der moralischen Begriffe
  (On the origin of moral terms)
- Gut und Böse nach der Lehre der Introgenese
  (Good and evil according to the teaching of Introgenesis)
- Similation und Bindung
  (Similation and Attachment)
- Über Philo und Ontoplasma. Zur Verlängerung des Lebens
  (On Philo and Ontoplasma. For the extension of life)
- Die Bedingtheit der physiologischen Äusserungen der Affekte
  (The relativity of physiological expressions of affects)
- Die Energetischen Vorgänge bei Epilepsie und die Möglichkeiten zu ihrer Beseitigung
  (The energetic process in epilepsy and the potential for its removal)
- Zur Frage der Beteiligung der trophischen Nerven in der Entstehung des Neoplasmas
  (On the question of the role of the trophical nerves in the formation of neoplasma)
- Wille und Vorstellung und ihre gegenseitige Abhängigkeit nach der Lehre über die Energetik der Psyche
  (Will and imagination and their interdependency according to the theory of the energetics of the psyche)
- Geist der Rasse in Abhängigkeit vom Klima (mit Berücksichtigung der eugenischen Frage der Cerebralentwicklung der Juden in Palästina)
  (The Mind of races/species in dependency on the climate (under consideration of the eugenic question of the cerebral development of the Jews in Palestine))
- Jüdische Nationale Bewegung vom Standpunkt der Psychologie des Unbewussten
  (Jewish National Movement from the perspective of the psychology of the subconscious)

- Atmung (Prana) und Entstehung des Gottesnamens
  (Breath (Prana) and the origin of God's name)
- Objekte als Mnemen der Vorstellung
  (Objects as mnemes of imagination)
- Raumbegriff bei den Zwangsneurotikern
  (The concept of space in obsessive-compulsive disorders)
- Zeitbegriff
  (Concept of time)
- Über die Möglichkeit der Entstehung der Anorganischen Welt aus der Organischen
  (On the possibility of the formation of the inorganic out of the organic world)
- Kunstwerte vom Standpunkte der Introgenese
  (Art-values from the perspective of introgenesis)
- Zur Frage des Zufalls
  (On the question of chance)
- Eine Analyse von Freuds Standpunkt in den Fragen der Parapsychologie
  (An analysis of Freud's perspective in questions of parapsychology)
- Sado-Masochismus und die homosexuellen Regungen in der christlichen Religion
  (Sadomasochism and homosexual tendencies in Christianity)
- Maggid und Kontrollgeist
  (Maggid and the spirit of control)
- Zur Behandlung der Szysotymer und Paranoiker
  (On treating Szysotym-disorders and paranoid disorders)
- Opferwilligkeit als Störung des introgenetischen Triebes
  (Self-sacrifice as a disorder of the introgenetic instinct)
- Die physiologische Bedeutung der menstrualen Ausscheidungen
  (The physiological meaning of the menstrual discharge)
- Über Entwicklung der Arten und der Lokalzentren
  (On the development of species and local centers)
- Zum Verständnis des Tbc-Prozesses. (Ion-Gehalt der Gewebe)
  (On the understanding of the tuberculosis process (Ion content of tissues))
- Brand. Alles oder Nichts als psychologisches Problem
  (Brand. All or nothing as a psychological problem)
- Television und Astronomie
  (Television and Astronomy)

The United States

In Cyprus

75th & Riverside Drive

The United States

In 1939, we were heading for the United States – but not without indecision. Much was at stake. My father understood that a war was about to begin and he was in conflict about where to have his family during a possible global conflict. We boarded the boat at Tel-Aviv, but when it anchored in Cyprus the next morning, we impulsively unboarded, perhaps because I had been sea-sick during the night – anything could have triggered my father's change of heart.

No sooner on the dock, the big boat behind us, regret set in – the boat had spat us onto the shore and sailed on. Traveling up the mountains of Cyprus by taxi, my father's fingers disturbingly fidgeted with something. Indecision and regret had set in. Many phone calls later, Cyprus forests and water falls going unnoticed, we boarded a boat destined for Trieste, Italy. (We had spent several summers in Cyprus riding horses, collecting acorns, and Nestle's chocolate bird pictures, never getting the one elusive card to complete the series for a big win.) By train we travelled to Cherbourg, France, and then on the Mauretania – on her maiden voyage – to New York.

On deck, winning a three dollar bet on a game horse race, my father thought that win may inspire me to gamble. Playing cards, "a symbol of decadence," and liquor were never allowed in our home.

In New York, the authorities prevented us from leaving the boat because my father had brought with him about $4,000 – a sum not permitted for temporary immigrants. We were required to stay on the boat overnight with the boat personnel. We stood on the deck like prisoners looking at the New York skyline. I thought we would be returned to Palestine; however, the next day, at Ellis Island – a depressing, unnatural detention of a conglomerate of humanity – my mother's brother had us released. Sigi, our "Uncle from America," who had on a visit to Palestine brought us dolls and "Patent Americai" (American Patents), had escaped matrimony until late in life, fathering two sons, the older ending his life with pills. We stayed overnight at Sigi's friend's house and the next day, my parents, my sister and I boarded a Fifth Avenue doubledecker bus. As it turned into Riverside Drive, my father directed us to get off. On the corner of Riverside Drive and 75th Street we rented a circular one room studio, where we all slept, cooked on a hot plate, and existed for months. We ate heated cans of Rokeach vegetable soup which Sigi, who worked for the company, provided. We did not take the Fifth Avenue bus again, although it stopped at our door. It cost a dime, a nickel more than the regular bus blocks away. Money ran out and a friend of my father's, Mendel Aviv, loaned us $100.

Aba
1939 World's Fair in New York

אתחלון גבא זלה צעורה, זלה צעורה
מא אבא
Nov 15. 67

"From us steps forward LaLa the courageous,
From Aba,
Nov. 15 1967"

Five weeks after we arrived in New York, World War II broke out. About the same time, my father learned that Freud died, just as he was about to send him his work on the reinterpretation of Freud's dreams.

For eight months my father conducted extensive library research on Greek legends and Egyptian writings which revealed much support for his Oedipus hypothesis. It was at this time that my father became friends with Professor Horace Kallen of the New School for Social Research. More than a decade later Kallen publicly supported and defended my father from Harlow Shapley of Harvard University, who led the scientific "inquisition" against my father for the next 30 years. Kallen was impressed with my father's work on that subject, and gave the manuscript to a publisher in New York. The publisher's serious interest in publishing this book changed our stay in the U.S. from a temporary one to a permanent one.

Freud and his Heroes was never published, but the part on Freud's dreams was published by Dr. Smith Jelliffe in the *Psychoanalytic Review* in October, 1941, as »The Dreams Freud Dreamed«. Not until two decades later my father enlarged the chapters on Oedipus and Akhnaton and they were more thoroughly documented and published as a book titled, *Oedipus and Akhnaton*, in 1960 by Doubleday.

I was enrolled in Julia Richman High School even though I had never attended 8th grade. Taking my sister for registration, my father decided to ask about me – He was afraid that if he enrolled me in a public school, I would be set all the way back since I did not know English. In a private high school, he reasoned, they could not set me back. The word I heard most often in the all-girls' high school was "cute". I was little and flat-chested, among high school girls who looked very sophisticated and all made up. For one entire year, I wore a checkered red skirt, a blouse and one dress.

I played the viola in the school orchestra and on one occasion sang an A to the piano tuner, misunderstanding the music teacher's request to ask the piano tuner for the correct A.

My mother read my book assignments at night. In the morning before school she filled me in on the cast of characters and the plot.

On one occasion, my father, all six feet two inches of him, appeared sternly in my school to reprimand a teacher who had poked fun at me. I had answered in Hebrew on an English grammar test. No one could get away with mistreating me. My father protected me from everyone.

In my teens, I bought a 15 cent tie from a street vendor – a grey tie with red diagonal stripes. I had no money to take the subway home, which then cost a nickel. I asked permission to ride the subway for free, which was granted. Upon telling my father the story, he was so impressed that he wore that tie to important occasions and told the story many times.

My father thought of me as courageous. He inscribed two of his books to me in the last years of his life with some reference to my courage, a quote from a children's game, – "... from us steps forward LaLa the courageous." My sister called me LaLa when I was born, not being able to pronounce "Yalda" ("girl" in Hebrew).

72nd & Riverside Drive

425 Riverside Drive (114th Street)

526 West 113th Street

In New York we moved several times.

In 72nd & Riverside Drive we had three rooms.

On the 8th floor of 526 West 113th Street we lived most of our years in New York. Our living room was converted into a ping-pong room where our friends would gather.

The final leg of our travels from Palestine occurred in the 1940's. We were still living in New York when my father arranged a visit to Einstein whom he knew from Germany. On that occasion my mother played chamber music with Einstein – quiet, stooped and sockless. To this day, every time I pass his house on Mercer Street, I think of the vegetable garden in the backyard, which he had shown to us.

Margo Einstein, a youngish-looking ancient woman whose art work and pleasantness were a joy, continued to live in the house into the 1980's. Miss Dukas, Einstein's life-long secretary, sharp faced and mean spirited, did not like the published reports that an open copy of my father's book, *Worlds in Collision*, was on Einstein's desk at the time of his death.

The executor of Einstein's papers resided across the street from my office in lower Manhattan. I got into a conversation with him in a neighborhood coffee shop, which led to getting permission to publish the letters Einstein had written to my father.

This visit to Einstein was our introduction to Princeton. The desire to be closer to Einstein encouraged my father to move one last time – to 78 Hartley Street, Princeton.

To me the trip from New York to Princeton had particular meaning. It was a safe trip. Traveling in Palestine had been fraught with danger. Bombs were thrown into buses and out of trains. I recall one woman lying on the sidewalk pleading, "I don't want to die," having been struck by a bomb thrown out of a train speeding through Tel-Aviv. My friend's father was shot and killed in Jaffa. My young cousin Raya, Aunt Trudel's daughter, who had just gotten married, was killed by a Syrian grenade thrown into a border kibbutz. On our visit to Israel in 1967, traveling from Tel-Aviv to Haifa by car and seeing Arab villages along the way, an early fear was ignited.

Raya

78 Hartley Avenue in Princeton

Old Porch on Hartley

The old house on Hartley Street in Princeton still stands, only recently having passed from the family hands. On the stone porch surrounded by a wrought iron railing that Aba had designed, I can still remember the two summer chairs whose green canvas bands sagged to match Aba's bodily contour. He sat there and contemplated, sometimes talking with visitors such as Lewis Greenberg, a devoted, ambitious man who brought Aba good tidings. In good times, the porch was a pleasant place to procrastinate. When depression set in, the misery of a man losing his self esteem and a wife who tried in vain to reverse the process was confined to the inside of the old stone house and the porch remained barren.

The house was built in the early part of the century. When he bought it in 1952, it had a white trellis, huge steps leading to the porch and clear fields across the street. The adjoining garden bloomed in the spring with white and pink flowers, and red roses climbed the wall facing the garden. Across Hartley was an empty field. Aba renovated the wrought iron fence and planned the garage and deck and two bedrooms in the back of the house. When the addition was partially complete, Aba found his association with what proved to be a dishonest builder intolerable, and the work came to a standstill. With the help of my son, Rafael, a capable architect and a compassionate grandson, I undertook to complete the job. My parents stayed at their beach house for several weeks while the work progressed. They returned to Hartley when the work was complete only to dislike the ceiling lamps.

In Princeton my parents each had a separate bedroom. My father went to sleep very early, sometimes as early as seven in the evening, and woke up at four in the morning and began to work. Early hours were my father's most creative. The New York Times was hidden from him with his knowledge until after he spent the morning hours creatively. My mother, on the other hand, stayed up till one or two in the morning, and liked to sleep later. She listened to the radio or television, and would sleep with it blaring all night. My parents protected each other's sleep never waking the other up. What kept them alive so long, I believe, was their eating habits of small meals and sleeping whenever they were tired. Every morning Ima was happy to wake up and told Aba that as long as nothing hurt they should count their blessings, which they often did. "Life is not life insurance," she said acknowledging her mortality. She particularly loved the fall colors, painted oil pictures of trees with orange leaves, and wrote a poem about it.

My parents ate substantial lunches and the rest of the day, ate lightly. Toward the end of his life, Aba became a finicky eater and my mother, who disliked cooking, tried to please him without success. He often stood gazing into the refrigerator contemplating what appealed to him.

He had always been helpful around the kitchen. My mother soaked the pots at night and my father washed them early in the morning when he got up. He washed his socks and underwear, and before buying a washer and dryer, he washed towels and sheets in the bathtub. When he shopped, he bought the biggest apples and pears. He always carried the groceries, my mother not liking to carry anything heavy. She even emptied her purse of quarters, they were heavy.

My father loved nuts and with his big hands cracked one nut against the other, an impressive feat. Aba had large hands with a little wart-like extension on each hand – I recall him saying that he had been born with six fingers on each hand.

Aba had difficulty coming to terms with new men introduced into his family. The difficulty wasn't just toward the men who wanted to marry his daughters. Doki, my niece's future husband, arrived from Israel on a foggy night to meet my parents. Coming down the stairs, bending his head to avoid the low ceiling over the last step, my father looked critically at him as he stood in the entrance room. Doki was forced to bear the uncomfortable stare.

My father could be quite forceful in establishing the terms by which men could enter our family. When I married Sidney (Sid) in 1946, in the taxi, on the way to my wedding, Aba announced that if Sid used his middle name "William", he would not permit us to get married. William was never used again.

When I announced that I wanted to marry Sid, my father researched his family, insisted that Sid get a medical checkup and then bought me a boat ticket to go to Israel to think it over, only to change his mind the last minute concerned that perhaps Sid was the love of my life and that later my father would be blamed for a breakup.

A seventeen guest wedding excluded Sid's religious family save his parents and aunt and uncle. Our small wedding was Aba's decision, and he gave us a monetary gift in place of a big wedding.

Before we married, Aba suggested Sid go to college, and made an agreement with Sid's parents to share the tuition. After the first year of study at NYU, my father undertook the tuition himself. He told Sid what courses to take, particularly history. Sid, relying on Aba who symbolically replaced his own father, got a degree in economics.

When I registered at NYU I did not tell my parents about it until the semester was over. After that, my father paid my tuition – but I still did not tell him what courses I was taking. As a child my father had frequently lectured to me, and I was glad he had replaced me with Sid. For years, I had been a captured audience.

Aba's concern about new men in the family extended to Shulamit's future husband as well. When he was about to arrive from Israel to marry her, my father wrestled with her choice of the right man. Exasperated, I accused Aba in Hebrew of having everything all mixed up in his brain, for which I was not easily forgiven.

Several years after our marriage, Sid and I drove cross country to visit his parents who had moved to Los Angeles. Sid wanted to remain. Having had a difficult time with my parent's illness, he had to complete his degree on the west coast. I managed to complete my master's degree at NYU for which my father expressed admiration. Sid persuaded me to remain in Los Angeles. My parents packed and shipped our belongings to us from the small studio apartment we had decorated on 97th Street.

Accepting our decision to stay in Los Angeles, my parents bought us a house and visited us. I was pregnant with our first child, Naomi. When we were first married, Aba told me to wait to have children. Five years into the marriage, he gave us "permission" to have children and a year later Naomi, at five pounds two ounces, was born. Not relating to babies, it wasn't until years later that Aba showed his love for Naomi.

Rafael was born two and a half years later. When he was two months old, we moved back to the east coast. My father expected a lot from him, and we often remarked that it was good my father did not have a son. Very proud of Rafael's architectural talents, he nevertheless was very demanding of a boy. When at one Seder my father was depressed and sat unresponsive, Rafael in particular, was sensitive to Aba's plight and suffered along with his grandfather.

Carmel was born in Princeton two years after Rafael.

My mother adored our three children, especially when they became adults. Carmel, whom she called Carmelli, patiently helped my mother to read after she had had the stroke. A special education teacher, Carmel was able to get my mother to read for her.

The Passover Seder had a particular meaning in our house. When in Princeton my father conducted the first Seder in our home. He explained his theories to the children in between reading the Haggadah. He paused often to elaborate on his ideas.

The fun part was when the children hid the matzoh, which he looked for but never found. Getting it back, he gave everyone at the table a gift – books, belts, pins, etc. He went shopping for the gifts on the day of the Seder, as though he did not have more important matters to attend to.

Reading the Haggadah he'd stop at a sentence praising those who tell the story of the Exodus. Pleased, he stressed that his books told of the Exodus.

When the wine made the children giddy, he simply went on reading, soon giving up himself.

The Seder started on a serious note and ended with joviality, and my father's singing a solo of "God has Built His Home in Peace". The song was particularly meaningful for the children. Aba, who liked to go to sleep at seven p.m. would sit on the couch as the children continued to sing. Rafael in particular would sing the "Ehad me Yodea?" ("One, Who Knows it?") with great musical skill.

Aba wanted to invite Einstein to our Passover Seder when the news came that Einstein had died. Many years earlier, when my father was about to send Freud his Reinterpretation of Freud's Dreams, Freud died. The paper dealt with Freud's unconscious desire to convert to Catholicism and the guilt he suffered – the persistent theme in Freud's dreams. These two recollections of Einstein and Freud point to my father's procrastination, but also to his approaching fields from a refreshing new angle.

Our family Seders provided another important link with Aba's legacy. He told us that what started him on his research was a Passover Seder where he noted the reference to jumping hills and the ten plagues. He thought perhaps a natural catastrophe had taken place, and if it was of great magnitude, ancient records from around the world would describe similar events. He went to the library, first to the 42nd Street library, and later to the Columbia University library, finding what he expected.

I remember Aba coming home from Columbia University library very excitedly as he began his research. Insights were coming daily, and after each trip to the library he talked to my mother sometimes for hours.

He was such a prodigious reader and had such a remarkable memory for the things that he read. With so much information at his fingertips, my father seemed to get pleasure in accumulating more. He bought himself an encyclopedia, and then bought a set for each grandchild. He spent many hours reclining on his bed, one arm supporting his head, his eyeglasses on his forehead, reading the encyclopedia, sometimes playing games with my mother, checking to see if he knew information before reading it.

When, on one occasion, I said to him, "penny for your thoughts," although it was mostly easy to "read" his thoughts by his facial expression, he answered, "I have no thoughts worth only a penny." He once calculated that the word "the" appeared so many times in the encyclopedia that it would fill one volume of a twenty-four volume set.

Aba typed his manuscripts and letters with the forefinger of each hand. His English was remarkably good and some critics said that he was guilty of writing many times better than his critics.

In New York he brought the manuscripts to Marion Kuhn, a woman ravaged by polio who barely moved on crutches. She was able to read my father's handwriting and was editorially capable. To the end of their lives my parents sent Miss Kuhn monetary gifts. Her note was taped on their bathroom wall near the mirror where it was read daily. In the note she counted her blessings, ending with "to ask for more would be a sin!"

Ima's Sculpture

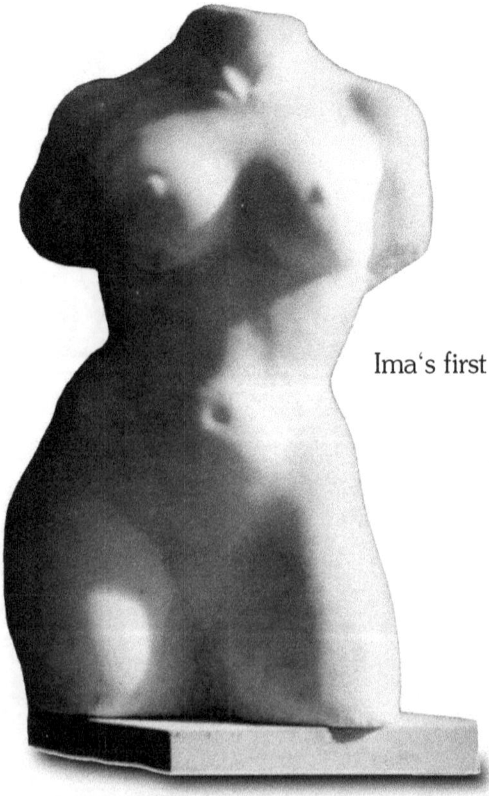

Ima's first work

Ima's Sculpture

While in New York, Ima studied sculpture under Maldarelli at Columbia University. A stone torso was her first work.

Her sculptures were exquisite, comparable to the works of the great masters.

They were included in annual shows at exhibitions of sculpture:

The New York Historical Society 1943
Whitney Museum 1946, 1947, 1955
Pennsylvania Academy of Fine Art 1952
Audubon Annual Exhibit 1946, 1952
Montclair Art Museum 1955, 1958

She repeatedly exhibited (by invitation) at the Whitney Museum and the Pennsylvania Academy of Fine Arts and was included in the Metropolitan Museum of Art (New York 1950). This was a National Competitors Sculpture Exhibition of over 5000 works submitted and 94 were accepted. Her sculpture "Caryatid" was the first of the 50 photographs in the catalogue of the Museum.

She received the first prize in Sculpture from the New Jersey State exhibition at the Montclair Museum, New Jersey 1958.

Caryatid

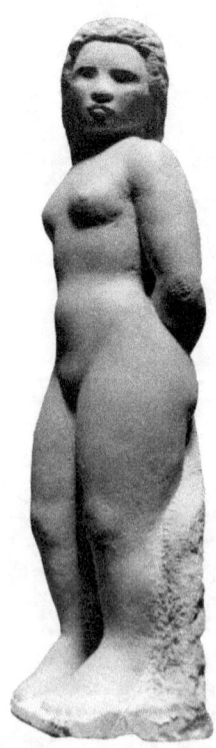

Ima's work is sensitive, sad, beautiful and refined.
Here are some reviews of her "Mother and Child" sculpture:

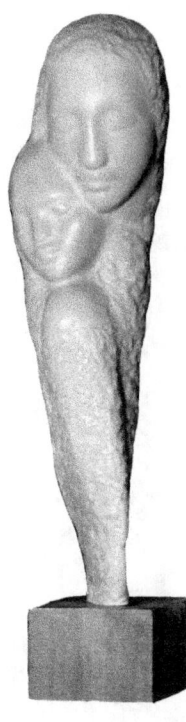

Mother and Child

Exhibition at The Whitney Museum of Art, New York

> A piece of particular charm is Elis Velikovsky's simple, sensitive "Mother and Child" (The Herald Tribune).

The photograph of "Mother and Child" was reproduced in Herald Tribune selected from the entire show.

Exhibition of the Audubon Artists, New York (National Academy)

> Sculpture seen ... at the Academy provides some very satisfying moments. Memorable among them is "Mother and Child" by Elis Velikovsky, an expertly handled fragment of stone, rich in content and suggestion. (The Art Digest)

One-man show, Princeton University, Wilcox Hall

> In "Mother and Child" we get a feeling of something growing out of this beautiful material. The shape of the work has its own beauty from every angle. The line and mass distribution plus the surface interest of rough and smooth give import to the subject." (The Packet, Princeton, N. J.)

Aba, Stone Head

Aba looking at Ima's self-protrait

Aba and Ima with her sculpture

Ima's Sculpture

My father loved my mother's sculptures. He did not let her sell any of her work for any price, and only agreed to gift copies.

He had his favorites. One was a lead head which, while my mother was hammering, he ordered her to stop pounding. He was afraid she would ruin the magnificent facial expression she had achieved. She nevertheless continued to hammer. My father, angered, got into his car and drove off. While driving around Princeton, the odometer turned a full set of numbers to the next thousand. My parents customarily held hands when such an event took place in the car. As the story was told, my father drove backwards all they way home so the odometer reversed itself, then asked my mother to join him on a ride, and when the numbers turned again, they held hands.

My father liked my art, too, except for what he labeled the "freaks". He was unhappy when I decided to study psychoanalysis. He wanted me to continue painting. Later, however, he approved of my new career. If he liked something, he had very strong opinions and he felt he was right. He did not talk to me for days when a mural of violins I had drawn on my living room wall, which he had admired, had gotten dirty and was covered over with white wall paint.

Six heads my mother created: two in stone, three in lead and one in flesh & blood of Ruth

Aba – The Observer

Lake Success—or Lake Failure?

Day of Judgment Opens in Shadow as Nations of World Debate Doom of Justice

By OBSERVER

IT IS LIKE a scene on the Day of Judgment. The time is the day after World War II, that had been fought on land, on sea, and in the air. A shadow lies over a desolate world, for already it is the twilight before the darkness of another World War that will eclipse the previous ones and may mark the end of the age of man on earth. The world with its two billion human beings sends emissaries from all its nations to its greatest metropolis. For a year and more they debate, and search, and argue for and against giving a little strip of land, 12 miles wide, to a stateless nation that lives there, the most ancient of them all, to be called home.

* * *

THE SUN rises and goes down; the streets are filled with people; cars run on winding highways; trains speed underground; and life goes on its way. But the sand runs low in the hour glass, and the weapons of destruction are piled high, and still the conscience of the world deliberates. To give the people of the Bible their Promised Land as agreed to by 55 nations at San Remo 28 years ago? To give them, perhaps, only the part that is this side of the Jordan? Or maybe only a strip twelve miles wide?

The nations of the world send emissaries from 12 of their number to investigate on the spot and to report. The emissaries return; the nations of the world again deliberate in commissions and vote, in committees and vote, in the plenum of the Assembly and vote. Finally, they appoint emissaries of five nations to give the narrow strip of land to this most ancient people.

The nations around the Holy Land move their bands there to destroy what Israel has built; and those on the isles of "The Ten Lost Tribes" (as the English say of themselves) send arms to the aggressors to make the destruction possible; and those in the land of the Star Spangled Banner put an embargo on arms needed by the ancient nation for the defense of its home.

* * *

THE NATIONS of the world reconvene. They are given a last chance to make good the evil which they and their fathers and their forefathers did to a homeless people, to wanderers over the face of the earth since the day they lost their home in a war of independence with Rome and through all the generations when they were persecuted for being true to their faith and to their heritage.

But the nations repent of their openhandedness. A twelve-mile strip? Too much! They were too generous! Return the judgment of the nations for reconsideration. Let us assemble together again at Lake Failure; it is certainly too much, a twelve-mile strip.

SAYS THE Prophet Isaiah (Chap. 43): "Let all the nations be gathered together, and let the people be assembled. ... O, Israel, Fear not: for I have redeemed thee. ... Fear not: for I am with thee. ... I will say to the north, Give up; and to the south, Keep not back; bring my sons from far, and my daughters from the ends of the earth."

In these days, in dark storerooms, missiles by the thousands are heaped, one of which sufficed to snuff out the breath of seventy thousand people of Hiroshima. Whoever created this world—or did it create itself?—man can destroy it.

If the nations of the world, Christian and Moslem and Buddhist alike, sitting in their Tribunal in this year 1948, will twist justice and empty it, and will stretch out their hand to extinguish the hope of the eternal people to return home, then:

"*Behold the nations are as a drop of a bucket, and are counted as the small dust of the balance ... All nations before Him are as nothing; and they are counted to Him less than nothing, and vanity*" (Isaiah 40).

April 16, 1948, reprinted with permission from the New York Post

In 1948 my father authored a weekly column for the New York Post about the Jewish problem. He used the name "Observer" and wrote about 40 articles.

Message to Lady Astor

Anti-Semitism Wanes as Israel Advances; Peeress Mistook Her Friends for All U. S.

By OBSERVER

Lady Astor, returning to England after a short stay in America, announced that she was startled to witness how much anti-Semitism had increased in the United States as a result of the Palestinian dispute. Fortunately, the Atlantic Ocean, which she crossed to bring this information to England, did not blush crimson. After all, the lady was not so much misinformer as misinformed herself, and the reason for this is that so many of those with whom she associates are Jew-baiters. In her innocence she decided that her thinning society is the American people.

The truth is that anti-Semitism is on the ebb in America. And the cause is—Israel. The noble and honest stand of the Jews in the United Nations in defense of their human rights has brought to their side most of the nations and the overwhelming majority of Americans. Their courageous fight in Palestine has evoked a spontaneous feeling of admiration among all from the bottom to the top of the American people.

Seven hundred thousand Israelis, constituting a two-weeks' old state amid the vast Arab area of the Middle East, fight against odds in numbers and odds in weapons with unsurpassed courage and an unbroken faith.

There are no Israeli refugees from Palestine. There are no draft dodgers in the nation. There are no deserters from the field of battle. These people fight with small arms against fieldguns; they go afoot against tanks; and when British fliers in Egyptian bombers or Egyptian fliers in British bombers bomb the Israeli capital, children bask in the sun on the capital's beach, paying little attention to the bombing.

"Hundreds of Legion shells have fallen on Jewish sections of Jerusalem for two weeks," Kenneth Bilby wires to the Herald Tribune. "The Jewish weakness, as the Arabs well know, is lack of heavy guns. No Jewish artillery has been fired within the Holy City. Their success in withstanding the Legion to date is a tribute to their tenacity and to the spirit of the individual soldier."

When, a year ago, Rabbi Korff, a New Yorker, hired in Paris a two-seater passenger plane to fly over London and drop leaflets of protest against the deportation of the immigrants of "Exodus 1947" from the shores of Palestine back to the German concentration camps, and the rumor spread that the Rev. Korff intended to bomb London, the morning papers carried dispatches from London saying that its population "became chilly with panic" when the radio announced that Rabbi's intention. This was despite the fact that simultaneously it was broadcast that Rabbi Korff had been arrested before the plane left the Paris aerodrome.

Apparently the nerves of the Londoners are anaphylactic to bombings since the days of the war. Anaphylaxis is a medical term for an exaggerated reaction to the second administration of a drug. But why does this story crop up in this column?

I remembered it when I read about the Israeli city of Tel Aviv's being under incessant bombing for five consecutive days, a city that had no fighters to intercept the planes and little other anti-aircraft protection, since only a week before the Israelis could not legally possess any weapons at all. Most persons in the cafes did not even lift their heads from their newspapers when bombs crashed into the streets. These Israelis—yesterday's Jews—also had plenty of reason to be anaphylactic to the danger of destruction. But instead of anaphylaxis, they had immunity in their hearts.

In his book on anti-Semitism in America, "A Mask for Privilege," Carey McWilliams writes: "It is notorious that the sadist persecutes the weak and defenseless not merely because it is safer but because it is somehow more pleasurable than to persecute the strong."

The epic of the Israeli fight for their homeland enchants all the peoples of the world, fills them with respect, and destroys anti-Semitism together with its pathological roots. The Israelis are not defenseless but instead deal blow for blow to seven armies of seven states. Even in Arab countries anti-Semitism is on the ebb. And in America, in its stead, a feeling grows against the Empire that cowardly hides itself behind the backs of Arab states.

The respected educator, Alvin Johnson, in a letter to the New York Times on June 1 wrote: "They (the British) imagine that the American policy of recognition of Israel is dictated by concern over the Jewish vote. It is not. It is dictated by concern over the American vote. Israel has ceased to be a Jewish issue. It is an American issue, the issue of republicanism against the imperialism which backs barbaric 'kings' and emirs. I do not find half the bitterness against England among Jews that I find among my solid Declaratio nof Independence Yankees of the Middle West."

This is the message that Lady Astor should have brought to the British shores.

June 4, 1948, reprinted with permission from the New York Post

A Mountain Was in Travail

2,500 Square Miles for Israel; 42,500 Square Miles for Abdullah

By OBSERVER

Sept. 28, 1948, reprinted with permission from the New York Post

A mountain was in travail. The United Nations, the so-called peace loving nations, the self-styled free nations of the world, for almost two years busied themselves with the Palestinian problem. Committees and commissions, the General Assembly and the Security Council, all labored heavily. This fall the mammoth organization swam over the ocean from the metropolis of the New World, New York, to the metropolis of the Old World, Paris, where it resumed its ponderous deliberations on Palestine.

The mountain was in travail. For twenty months great nations, world powers, exalted statesmen, shrewd diplomats, presidents of states made declarations before their nations, before parliaments and congresses, to the press and on the air, concerning Palestine, its partition, the granting of statehood to the most ancient nation on earth. If the ticker tape with all these words were stretched in a straight line, it would reach the moon. If the pages of the articles and books that were published on this subject, and the cables and wires that were sent, were placed side by side, they would extend to the sun.

Then came the time for delivery in this great travail. The Secretary of State of the mightiest country in the world stepped forward and mounted the rostrum to address the delegates of fifty-eight nations. The press of the world and the radio carried his words; and millions, indeed hundreds of millions of people read them and listened to them in all parts of the world— in the Western Hemisphere, in the Eastern Hemisphere, people of the white race, the yellow race, the black race.

Secretary of State Marshall supports the Bernadotte plan. This plan is actually the third partition of Palestine. The first partition was executed when Great Britain, without authorization, severed Transjordan—more than three-quarters of the land— from the territory of Palestine mandated to her by the 51 nations of the League of Nations for the creation of a National Home for the Jews. This area Britain gave to an Arab emir, Abdullah, with the understanding that the land this side of Jordan should serve as the National Home.

The second partition took place with the vote of the United Nations Assembly on Nov. 29, 1947. The 11-nation commission that had been sent by the United Nations to investigate the Palestinian problem proposed that the remaining part of Palestine be divided between the Arabs and the Jews. The Jews were to receive about half, including the entire Negev; but before this proposal came to a vote in the General Assembly, part of the Negev was cut off and transferred to the Arab portion of Palestine. Under pressure of the desperate need of Jewish migrants from Europe, the Jewish community in Palestine, with the exception of the dissidents, accepted this sacrifice also.

The third partition is in the making with the late Count Bernadotte's recommendation that the Negev be amputated—a recommendation so generous that it chops off with the Negev a part of the land of Judah.

What, then, remains? What do the nations clamor? About what do the papers print headlines? It is the hour of delivery for the mountain that was in travail.

* * *

What remains is 2,500 square miles for the State of Israel. The State of New York can hold 21 such states as they offer for Israel; California has room for 63; Texas for 107 states of that size.

The Secretary of State is very generous. The President of the United States is exceedingly generous. The delegates of the nations are magnanimous. The freedom-loving nations of the world have an open hand.

Look at your common "gift." You gave it with a feeling of greatness of mind and elevation of soul, these 2,500 square miles from an original 45,000, or from the more than 10,000 square miles on this side of the Jordan. To Abdullah—42,500 square miles; and to Israel—2,500 square miles. You "gave" it. The boys and girls of Israel fought for it, for the Negev and for Jerusalem, too; on the beaches, on the roads, in the hills, on the streets, and bled and died. You gave them not one single gun to defend themselves.

Bells ring triumphantly in our soul; the sky is filled with angels rejoicing at our bounty and munificence. We have indemnified the martyred nation; we made good our bigotry, our negligence, our callousness.

The Battle

Photo by Stuart Crump Jr.

The Battle

While conducting library research for his book on Freud and his heroes, my father became intrigued with articles about the geological formations around the Dead Sea. A friend pointed out to him that, according to the book of Genesis, the Dead Sea was a plain in the days of Abraham. Only when Moses and Joshua arrived there after the Israelites fled Egypt was a body of water found to exist. This raised questions in my father's mind: Was the Dead Sea formed in the days of the Exodus, the result of a natural catastrophe? Was this catastrophe also experienced in Egypt? When he read the texts of an ancient Egyptian papyrus, he found that the answer was yes. In fact, the descriptions of a catastrophe in the papyrus looked just like a copy of Exodus. The same held true when he examined the ancient texts of the Chinese, the Maya, the Hindus, and the Babylonians, and others. For the next three years he went to numerous New York City public and university libraries, painstakingly gathering evidence that all pointed to the same conclusion, that a global catastrophe took place in recent historical times, and that it was triggered by cosmic collisions in space. These findings, he knew, would have serious implications for the fields of astronomy, geology, physics and anthropology. This discovery marked the beginning of my father's reconstruction of ancient history and his interdisciplinary theory of the Earth and the cosmos.

During the next several years, 1943 – 1950, my father shared his findings with scholars and scientists in different disciplines. Some scholars were open to my father's theories and of great help to him, such as Egyptologist, Dr. Walter Federn, son of a well-known psychoanalyst and family friend, Dr. Paul Federn. Another, Professor Robert Pfeiffer of Harvard University, was critical, yet supportive, and encouraged my father to publish his findings. Horace Kallen, Dean of the Graduate Faculty of the New School for Social Research, who read my father's manuscripts, wrote to my father:

> The vigor of the scientific imagination that you show, the boldness of your construction and the range of your inquiry and information fill me with admiration. (»The Censorship of Velikovsky's Interdisciplinary Synthesis« Lynn E. Rose, in *Velikovksy Reconsidered*, Pensee, New York, 1976, p.14)

The early years of research and writing were joyous years to my father. I remember him coming home from the forty second street library and later from Columbia University Library, and for hours discussing his findings with my mother.

In April, 1946, my father attended a talk given by Harlow Shapley of the Harvard College Observatory. To test his theory of cosmic collisions and global catastrophe, he needed an analysis of the atmospheres of Venus and Mars. Shapley, it occurred to him, could complete these tests. He approached Shapley, asked him to read the manuscript and to consider doing two spectroscopic analyses. Shapley agreed to this provided that Horace Kallen, Shapley's friend, also read and recommended it. Of course, Kallen was already familiar with the book and encouraged its publication. Despite Kallen's recommendation in writing, Shapley refused to read the manuscript and conduct the analyses, saying my father's claims were sensational. This was only the beginning of a life-long attack by the scientific establishment of my father's work and Shapley would become the leader of the pack.

Two months later, my father started his search for a publisher. One month after that, in July, he sent his manuscript to John O'Neill, science editor of the *New York Herald Tribune*, who, soon after, wrote of my father's theory in an article that appeared in that paper on August 11, 1946. By May, 1947, after several publishers turned down the publication of my father's book Macmillan Company agreed to publish *Worlds in Collision*, a contract was signed, and my father began the task of completing the final draft. By February, 1949, the galleys were ready, and the book would be released in April of that year.

Before the publication of *Worlds in Collision*, Frederick L. Allen, editor-in-chief of *Harper's Magazine*, was authorized by Macmillan to present an article summarizing my father's book. This was written by Eric Larrabee, an editor at *Harper's* and was titled *The Day the Sun Stood Still*. It appeared in the January, 1950 issue of the magazine and sold out in just a few days, but it provoked many angry attacks and protest letters to *Harper's* from scientists around the country.

Only a few days after this article appeared, a Japanese astronomer observed the first cosmic collision in modern times, between Mars and a planetoid.

One of the attacks was from Harlow Shapley, the astronomer from Harvard University who refused to conduct the spectroscopic tests. He wrote two letters to James Putnam, my father's editor at Macmillan. In one letter he said that he, other scientists, and the president of Harvard University were "astonished that the great Macmillan Company, famous for its scientific publications, would venture into the Black Arts without rather careful refereeing of the manuscript." Both Putnam and George Brett responded to Shapley's letters, Brett saying that opinions from three scholars would be sought before publication. Two of these scientists recommended it, one did not, hence, *Worlds in Collision* would be published. Professor C. W. van der Merwe, Chairman of the Department of Physics at New York University, was one of the three who read the manuscript and approved the book for publication.

When Shapley did not get the results he hoped for from Macmillan, he changed his strategy. He wrote a letter to Ted O. Thackrey, previously chief editor of the *New York Post* and then publisher of the *Compass*, a progressive newspaper. Thackrey had recently reprinted Eric Larrabee's *Harper's* article. He also wrote a favorable editorial about my father's work and its value to science. In his letter to Thackrey, Shapley called *Worlds in Collision* "the most successful fraud that has been perpetrated on leading American publications." He also asked Thackrey to print a refutation of my father's work, an article written by his astronomer-colleague, Cecilia Payne-Gaposchkin, a copy of which he attached to the letter. It turns out that Shapley had also distributed this mimeographed article to many members of the scientific community around the country.

Thackrey refused Shapley's request, but Payne-Gaposchkin's article, *Nonsense, Dr. Velikovsky*, was later printed in *The Reporter*. About the same time, Shapley recruited five scholars in different fields to denounce my father's theories in his own publication *Science News Letter*. And, a month later, he cited Payne-Gaposchkin's article in *Science News Letter*, urging all scientists to read her "scientific" refutation of *Worlds in Collision*.

Shapley, Payne-Gaposchkin, and Shapley's five critics from different fields did all of this without ever having read *Worlds in Collision*, which had not yet been published. These denunciations of my father's work were based on Payne-Gaposchkin's unscientific critique, which was only based on Larrabee's article in *Harper's*.

The efforts at suppressing my father's theories did not wane. In *Stargazers and Gravediggers*, my father tells of a resident of New York City who wrote to him saying he should

> ... prepare for at least a ten-year siege by entrenched bigots. During that time you will come to understand why Copernicus and others have waited until the last moment before breaking into public proclamation of their findings. The siege, you will find, will not be limited to yourself but will also menace your family. (*Stargazers and Gravediggers*, File II, »With Hatbrims Pulled Down«)

But, the siege did not end after ten years. It continued for the next fifty plus years to the present day. Despite Shapley having been a friend, Thackrey responded sharply and frankly to Shapley's letter, expressing his shock

> ... at the epithets you have seen fit to use in characterizing Dr. Velikovsky, a man of unusual integrity and scholarship, whose painstaking approach to scientific theory is at least a match for your own. ...
>
> You further suggest that, evidently through your efforts, there is now some question about whether Macmillan will go through with the publication, thus not only confessing to do direct damage, but to provide some evidence of having successfully damaged Dr. Velikovsky's work. ...
>
> It seems to me that you are making both a personal and professional mistake – a gravely serious and dangerous one – by the totally unscientific and viciously emotional character of your attack upon Dr. Velikovsky and his work.
>
> I am writing this advisedly, since it is obvious that you have seen fit to unleash a series of attacks, by no means directed to me alone, both against Dr. Velikovsky and against his work without ever once having taken the trouble to examine his work or even to glance at the evidential research with which it has been accompanied.

> I submit that, at the time of writing your letter, you had neither read the manuscript of Dr. Velikovsky's *Worlds in Collision*, nor a single piece of evidence in its support. At the most, it is possible that you had examined superficially a popularization of a very small portion of this work by Eric Larrabee of Harper's Magazine. (*Stargazers and Gravediggers*, File I, »Somebody Has Done You Dirt«)

And, regarding the article by Cecilia Payne-Gaposchkin, Thackrey characterized the article as "an attack upon a book which the writer has not read," and that

> ...the article attributes to Dr. Velikovsky statements which are not made either by him or in his manuscript, and then proceeds to quarrel with those statements as though they were authentic. This is, to say the least, a most unscientific method of criticism ... (*Stargazers and Gravediggers*, File I, »Somebody Has Done You Dirt«)

Setting up and knocking down strawmen would prove to be yet another tactic used repeatedly by scientists to "disprove" my father's theories. Misquoting my father's work was a pattern that started before the publication of *Worlds in Collision*, continued with his subsequent books and has persisted for more than half a century.

On March 10, 1950, my father waited for my mother to arrive at the Macmillan building, as he did not wish to view the first copies of *Worlds in Collision* unless she were there to share that moment. After her arrival, Macmillan employees from every floor in the building appeared, asking my father to sign their copies of his book (These books were not gratis, but purchased by the employees themselves.) The publicity director told my father that she had never before seen so many employees come to have their books autographed by an author. At that time, however, they also learned that the Museum of Natural History had returned their complimentary copies and that a sky show based on *Worlds in Collision* that was planned by Gordon Atwater, head of the Hayden Planetarium, would not take place.

Apparently, the opposition, led by Harlow Shapley, had escalated its war against my father. Otto Struve, director of the Yerkes Observatory at the University of Chicago and ex-president of the American Astronomical Society, joined the letter-writing campaign. He wrote to John O'Neill, science editor of the *Herald Tribune*, and to Gordon Atwater, also chairman of the Department of Astronomy at the Museum of Natural History. He asked that they cease their support of Velikovsky and his work. Both refused to comply with Struve's request. Atwater replied that although he did not agree with everything my father wrote, he believed it had great value for the scientific world.

Three weeks later Atwater was fired with one day's notice. However, an article he wrote still appeared in the April 2nd issue of *This Week*, a magazine published by the *Herald Tribune*. Although the editors felt the pressure not to print it, John O'Neill, after showing them Payne-Gaposchkin's mimeographed article, advised them to print Atwater's article. In it Atwater wrote that *Worlds in Collision*

> will have an explosive effect in the world of science. ...

and

> while *Worlds in Collision* is being condemned by large numbers of professional scientists, many other groups will welcome the book as a broadening influence in scientific, religious and philosophical fields (*Stargazers and Gravediggers*, File I, »The Book Is Launched, Atwater Thrown Overboard«)

Thirty years later Atwater was asked if he had any regrets about his association with Velikovsky to which he replied,

> I regret the way they treated Dr. Velikovsky. He was a wonderful man, and what they did to him was a disgrace. That's what hurt me the most. (*Stargazers and Gravediggers*, File II, »A Second Man Thrown Overboard«)

Atwater and Aba

O'Neill had also composed a review of *Worlds in Collision* for the same April 2nd issue (the day before publication and release of the book), but it was replaced with a review by none other than Otto Struve. Instead of presenting an objective, scientific critical analysis of *Worlds in Collision*, Struve offered an unscientific rejection of it with not a shred of data to support that rejection. His excuse was that because *Worlds in Collision* was not a book of science he did not have to analyze it on the basis of scientific terms (This tactic, to dismiss Velikovsky's work as unscientific, has been continually used by the scientific establishment until the present day, most likely to keep my father's theories outside of the realm of science, where they might actually be discovered to be accurate). Struve, instead, chose to cite what he considered to be a scientific work, Cecilia Payne-Gaposchkin's article in *The Reporter*, which misrepresented my father's arguments and contained many inaccuracies, not surprisingly, because she had never read the book.

Waldemar Kaempffert, chief science editor of *The New York Times* claimed to have read the book yet his negative book review in the *New York Times Book Review* contained facts that were present in the Payne-Gaposchkin article, not in *Worlds in Collision*. He too, it seemed, had relied on secondary sources for his review. Other establishment scientists joined the collective campaign to suppress Velikovsky's book all condemning *Worlds in Collision* in book reviews, many that were syndicated and printed in newspapers that reached all corners of the country. The names changed: Harrison Brown, atomic scientist, Paul Herget, Director of the Observatory at the University of Cincinnati, Frank S. Hogg, Director of David Dunlop Observatory, University of Toronto, but the reviews essentially remained the same: Payne-Gaposchkin's arguments were mimicked again and again complete with misquotes of my father's book.

Ralph Juergens cites numerous examples of these misquotes in his article, »Minds in Chaos«; printed in *The Velikovsky Affair: Scientism vs. Science*, edited by Alfred de Grazia, Ralph E. Juergens and Livio C. Stecchini, University Books, New Hyde Park, New York, 1966.

Here is one of many examples:

> Ignoring Velikovsky's alternate explanation that, perhaps in the grip of an alien magnetic field, a "tilting of the (earth's) axis could produce the visual effect of a retrogressing or arrested sun," Frank K. Edmondson, director of Goethe Link Observatory, University of Indiana, wrote: "Velikovsky is not bothered by the elementary fact that if the earth were stopped, inertia would cause Joshua and his companions to fly off into space with a speed of nine hundred miles an hour."
> This argument, first formulated by Gaposchkin, is at best disingenuous, for the all-important time factor, the rate of deceleration is completely ignored.

Despite these negative reviews, the public loved *Worlds in Collision*, and it remained on the best-seller list of the *New York Times* and the *Herald Tribune*. However, on

The Battle

May 25, 1950, the pressure from the Shapleyites was mounting on George Brett, president of Macmillan Publishing. He asked that my father come to meet with him. At that meeting Brett expressed his concern over the violent opposition from the members of the scientific establishment and the threat of a boycott of Macmillan's textbooks (Their textbook sales comprised 70% of total revenue). He said that never in his thirty-three years in publishing did he have to ask an author to release his publishing company from a contract. Then, he asked my father to do just this. He explained that he had found another publisher, Doubleday Company, that agreed to publish *Worlds in Collision*. After careful consideration, my father agreed to go with Doubleday. Soon after the transfer took place, James Putnam, my father's editor, was terminated by Macmillan.

IMMANUEL VELIKOVSKY

and

THE MACMILLAN COMPANY.

MEMORANDUM OF AGREEMENT

Dated June 7th, 1950.

MEMORANDUM OF AGREEMENT made this 7th day of June, 1950, by and between IMMANUEL VELIKOVSKY (hereinafter called the "Author"), party of the first part, and THE MACMILLAN COMPANY (hereinafter called the "Publisher"), party of the second part.

WHEREAS the parties hereto, under an agreement dated May 8, 1947, entered into a contract for the publication of a work entitled WORLDS IN COLLISION, which work was duly published on April 3, 1950; and

WHEREAS it is deemed mutually desirable that the rights in said work be returned to the Author by the Publisher and that he be left free to arrange for its publication elsewhere,

NOW, THEREFORE, in consideration of the covenants and agreements herein contained and of other good and valuable considerations, the receipt of which is hereby acknowledged, it is mutually agreed as follows:

1. The Macmillan Company hereby sells, assigns and transfers to Immanuel Velikovsky all its right, title and interest of every kind and nature in and to the work entitled WORLDS IN COLLISION, which rights it received under its contract with the Author dated May 8, 1947, as well as the plates of said work now owned by The Macmillan Company.

2. The Macmillan Company herewith tenders to the Author all copyright certificates pertaining to said work, it being its intention to divest itself of all interests under said contract of May 8, 1947, and to permit the Author to arrange for its publication elsewhere.

3. The Macmillan Company agrees that upon the signature of this agreement by the Author he shall be free to deal with said property, WORLDS IN COLLISION, as well as the plates thereof, as he may deem desirable.

4. The Macmillan Company agrees that within thirty (30) days of signature of this agreement it will render and settle to the said Immanuel Velikovsky accounts of all sales of said work by it in accordance with the provisions of said agreement of May 8, 1947, and the Author agrees that upon such payment all claims of any kind and nature against the said The Macmillan Company shall be thereby discharged.

5. The Author agrees that upon signature of this agreement he accepts the return of all publishing rights in said work from the Publisher to him together with the plates of said work and fully releases and discharges the Publisher from all claims of any kind and nature which against it he might, could or would have, excepting only that he retains a claim for royalty earnings accumulated and not paid to him by the Publisher until such time as accounts and payments have been rendered him.

The provisions of this agreement shall inure to and bind the heirs, administrators, executors, successors and assigns of the respective parties.

IN WITNESS WHEREOF the parties have hereunto set their hands and seal the day and year first above written.

Immanuel Velikovsky
Immanuel Velikovsky

THE MACMILLAN COMPANY
By *George P. Brett*
President

The Battle

Years later, Harold Latham, chief editor of Macmillan Publishing would write, in his book *My Life in Publishing*, that he always felt regret over Brett's decision. In 1965, Latham wrote a letter to my father saying that

> I remember well the commotion caused by *Worlds in Collision* and I do not remember with any pleasure the part that Macmillan played in the episode. I always felt that we made a mistake in taking so seriously the criticisms and demands of the scientists and textbook authors.
> I should have preferred to stand our ground and face our detractors and I think they might very soon have been put to the rout. But the decision was not mine to make. (*Stargazers and Gravediggers*, File II, »A Second Man Thrown Overboard«)

Two months after going with Doubleday Company, *Worlds in Collision* was published in Great Britain, where British scientists reviewed it in a similar manner, extensively misquoting my father. Juergens, in *Minds in Chaos*, describes how evolutionist J. B. S. Haldane, author of *Science and Ethics*, reviewed the book in the *New Statesman and Nation* on November 11, 1950.

> Haldane misquoted Velikovsky, then ridiculed the misquotation; he mismatched dates and events Velikovsky had associated with them; he concluded that the book was 'equally a degradation of science and religion.' (p. 32)

Until this time, scientists simply refused to read the book, dismissing it as nonscience, and misquoted it repeatedly. But, now a new strategy was being employed to discredit my father. They attacked his research methods, accusing him of misrepresenting his sources. This new attack was led by Cecilia Payne-Gaposchkin in an article in *Popular Astronomy* (June, 1950) and Professor Otto Neugebauer of Brown University in his article in the journal *Isis* (Vol. 41, 1950). According to Juergens in *Minds in Chaos*, Professor Neugebauer

> ... accused Velikovsky of willfully tailoring quoted source material.

To support this charge, Neugebauer specified that

> Velikovsky had substituted the figure 33 14' for the correct value, 3 14', in a quotation from the work of another scholar.
> When Velikovsky protested in a letter to the late George Sarton, then editor of *Isis*, that the figure given in his book was correct and the 33 14' was in fact Neugebauer's own insertion, not his, Neugebauer dismissed the incident as a 'simple misprint of no concern' that did not invalidate his appraisal of Velikovsky's methods. And the reprint was circulated by an interested group long after its errors had been pointed out. (p. 27-28)

The suppression of *Worlds in Collision* was a collective, organized one. Shapley later denied his role in it, but the following quotes, printed in Horace Kallen's *Shapley, Velikovsky, and Scientific Spirit* (In *Velikovsky Reconsidered*, Doubleday Company, Garden City, New York, 1976), speak for themselves:

On January 25, 1950, Shapley wrote:

> It will be interesting a year from now to hear from you as to whether or not the reputation of the Macmillan Company is damaged by the publication of *Worlds in Collision* ... Naturally you can see that I am interested in your experiment. And frankly, unless you can assure me that you have done things like this frequently in the past without damage, the publication must cut me off from the Macmillan Company. (Harlow Shapley in a letter (January 25, 1950) to James Putnam of the Macmillan Company)

Eight months later, Shapley said:

> The claim that Dr. Velikovsky's book is being suppressed is nothing but a publicity promotion stunt ... Several attempts have been made to link such a move to stop the book's publication to some organization or to the Harvard Observatory. This idea is absolutely false. (Harlow Shapley in a statement to the *Harvard Crimson*, printed in the *Crimson* on September 25, 1950)

The relentless attempts at suppressing my father's work did not stop after Macmillan Company transferred *Worlds in Collision* to Doubleday Company. On June 30, 1950, Fred Whipple, Shapley's successor as director at the Harvard Observatory, wrote to Eunice Stevens, associate editor of the Blakiston Company (owned by Doubleday) saying:

> ... I believe that the Blakiston Company is owned by the Doubleday Company, which controls its policies as well as the distribution of its books. I am now then a fellow author of the Doubleday Company along with Velikovsky. My natural inclination, were it possible, is to take *Earth, Moon and Planets* off the market and find a publisher who is not associated with one who has such a lacuna in its publication ethics. This is not possible, however, so the next best that I can do is to turn over future royalty checks to the Boston Community Fund and to let *Earth, Moon and Planets* die of senescence. In other words, there will be no revision of *Earth, Moon and Planets* forthcoming so long as Doubleday owns Blakiston, controls its policies and publishes *Worlds in Collision*. (Kallen, 1976, p. 25)

Twenty years rater, on July 2, 1970, Whipple denied this when he told Clark Whelton of *The Village Voice*:

> With regard to Mr. Velikovksy's *Worlds in Collision* there is no change in my attitude or in the situation since the book was first released nearly a decade (sic) ago. There is no truth to allegations that I sought to dissuade the Doubleday Company from publishing this book or any other book. ... (Kallen, 1976, p. 25)

The Battle

The vicious, emotional attacks on my father were relentless through the 1950's, 1960's and 1970's, with scientific and philosophical societies holding meetings for the express purpose of misrepresenting my father's theories and then debunking those misrepresentations. The conduct at these meetings was not at all serious and scholarly. Instead, their members met to mock and humiliate my father, who attended many of these meetings with the intent of participating in a serious debate.

One of these meetings was held by the American Philosophical Society in Philadelphia in April, 1952. One paper was delivered by Cecilia Payne-Gaposchkin, which repeated all of her previous attacks, based on misquotations, erroneous data and false accusations about my father's research methods. Other papers followed by geologists, archaeologists, and astronomers, all recruited to refute *Worlds in Collision*. My father was present at the meeting, which was a surprise to its members since he was not invited to present his side. He was, however, given time to present a rebuttal, which drew a long, enthusiastic applause. However, when he asked that his rebuttal be printed along with the other papers in the society's *Proceedings*, his request was denied.

First, scientists tried to prevent the public from learning about my father's theories by stopping the publication of *Worlds in Collision*. Now it appeared that they had devised a way to prevent the academics from learning about my father's theories by not publishing his rebuttals. In the fall of 1950, established scientists were given a chance to debate my father in *Harper's Magazine*. Harlow Shapley, Otto Neugebauer and others declined, but Princeton University astrophysicist, John Q. Stewart, agreed. But, predictably, he also based his criticism on Payne-Gaposchkin's *Reporter* article. Stewart's inaccuracies, however, were exposed in my father's reply. Finally, in the 1951 issue of *Harper's*, one establishment scientist, Julius S. Miller, professor of physics and mathematics at Dillard University, wrote a letter to the editor. In this letter he "cited what he called a 'glaring paucity and barren weakness of explicit criticism' on the part of Velikovksy's critics. He concluded:

(1) The Velikovsky notions are not altogether untenable,

and

(2) ... not yet refuted. (Juergens,1966, p. 34)

At that time, a few establishment scientists engaged in serious debates with my father and gave him the constructive criticism that he welcomed as part of scientific inquiry.

Walter S. Adams, director of Mt. Wilson and Mt. Palomar Observatories was one of these scientists. Even though he disagreed with my father's theories about the role of electromagnetism in space, he spent a considerable amount of time corresponding with him and even met with him at the solar observatory in Pasadena to discuss further the problems of celestial mechanics.

One time he answered one of my father's letters saying,

> I have tried to be quite objective in this letter since I dislike some of the almost abusive criticisms which have been written about your book. They are uncalled for no matter how strongly their writers feel on the subject.' (*Stargazers and Gravediggers*, File III, »Pursuing a Ray of Light«)

In 1952, my father published his next book, *Ages in Chaos*, which documented his reconstruction of ancient history. Professor Robert Pfeiffer, Chairman of the Department of Semitic Languages and Curator of the Semitic Museum at Harvard University, endorsed *Ages in Chaos* and authorized my father to include his comments on the dust jacket. But, William Albright, Spence Professor of Semitic Language at Johns Hopkins University, wrote a negative review of the book in the *Herald Tribune*, yet presented no concrete argument backed up by accurate historical facts. While many ancient historians were outraged at my father's reconstructions, none of these critics presented in the scientific press any scholarly analysis of his work. And, what of astronomer, Cecilia Payne-Gaposchkin, who had fashioned a career of defaming Velikovsky? When she heard that *Ages in Chaos* was released, she stated that she would read it, but was sure it would be as wrong as *Worlds in Collsion*.

When my parents moved from New York City to Princeton in 1952, my father met a number of scientists from Princeton University and was given opportunities to engage in serious scientific debates at the university. Once in 1953, he addressed the Graduate College Forum at Princeton about his theories in light of the recent finds in archaeology, geology and astronomy. At that talk he hypothesized that the earth's magnetic field extended as far as the moon and predicted that Jupiter emitted radio noises, just two of his advance claims that were later confirmed by space probes in April, 1955.

It was also during this time that my father became acquainted with Harry H. Hess, head of the geology department at Princeton, and Lloyd Motz of Columbia University. They often disagreed with my father's theories. Nevertheless, they still urged scientific associations to take my father's work seriously, especially after the space probes confirmed many of my father's advance claims. But these attempts were fruitless.

In 1955, my father's book *Earth in Upheaval* was published. This marked a new pattern of abuse by the scientific community to suppress Velikovsky's work. Instead of collective attacks, there was collective silence. The response by the scientific journals and reviewers was to ignore its presence. In the March, 1956 issue of *Scientific American*, Harrison Brown presented what was supposed to be a review of *Earth in Upheaval* but was actually a public scolding of scientists' behavior six years earlier with respect to *Worlds in Collision*. The review essentially ignored *Earth in Upheaval*, briefly dismissing it without any concrete evidence to support this dismissal. There were some favorable reactions to the book, namely those of the participants in a radio program, which included host, Clifton Fadiman, the Dean of the Graduate Faculties at Columbia University, Jacques Barzun, and Alfred Goldsmith, president of the Radio Engineers of America. All three praised the scholarship of *Earth in Upheaval* and agreed that the work deserved serious consideration from the scientific community. At the same time, Claude Schaeffer, professor at Collège de France and the archaeologist at Ras Shamra in Syria, wrote to my father saying that his own findings confirmed those in *Earth in Upheaval*, that natural catastrophes were experienced by Middle Eastern civilizations according to my father's timetable. He and my father corresponded for many years and in 1957 later met in Europe.

My father's next book *Oedipus and Akhnaton*, was also ignored by the scholarly community. Yet one of his critics, William Albright, couldn't pass up another opportunity to discredit him. He pronounced Velikovsky wrong again, this time on the grounds that early cultural contact between Egypt and Greece was an impossibility. Albright ignored or wasn't aware that there existed archaeological proof of the existence of Mycenaean artifacts in the city of Akhnaton.

On a positive note, though, world-renowned classicist, Professor Gertrude E. Smith of University of Chicago, reviewed it favorably in the *Chicago Tribune*.

During the late 1950's and early 1960's, my father continued addressing the faculty and students of the geology department at Princeton University at the invitation of Professor Harry Hess, and *Earth in Upheaval* was required reading for his geology course there. When the Mariner II space probe revealed that Venus indeed had a high surface temperature, just as he predicted years earlier, my father, feeling hopeful, submitted a paper to *Science*. It was returned promptly by the editor who never even read it, and an abusive letter was published instead from a Poul Anderson, saying that

> the accidental presence of one or two good apples does not redeem a spoiled barrelful. (Juergens, 1966, p. 45)

In September, 1963, the *American Behavioral Scientist* published three papers dealing with the Velikovsky controversy and the politics of science. The response from the scholarly community to this issue was a favorable one. Soon after, Eric Larrabee, who had introduced *Worlds in Collision* to the public via his article in *Harper's* thirteen years prior, wrote another article for *Harper's* about recent scientific discoveries that supported Velikovsky's theories. In this article, titled *Scientists in Collision*, Larrabee wrote that

> Science itself, even while most scientists have considered his case to be closed, has been heading in Velikovsky's direction. Proposals which seemed so shocking when he made them are now commonplace. ... His dismissal and suppression by the scientific community require of scientists an act of agonizing reappraisal. (*Harper's Magazine*, August, 1963)

Apparently, the pattern of emotional outbursts typical of the Harvard College Observatory resurfaced. The director at that time, Donald Menzel, was so outraged by Larrabee's article, that he wrote a reply to the article that was so abusive that the editor struck the following sentence:

> Velikovsky has been as completely discredited as was Dr. Brinkley of the goat-gland era or the thousands whom the American Medical Association has exposed as quacks, preying on human misery, by purveying nostrums or devices of no beneficial value whatever. (quoted in »Aftermath to Exposure« by Ralph Juergens in *The Velikovsky Affair*, 1966, p. 52).

This illustrates the relentless abuse that my father suffered for thirty years of his life. The Larrabee article, nevertheless, did generate support through letters to the editor, for example, from V. A. Bailey, Emeritus Professor of Physics at the University of Sydney and Lloyd Motz of Columbia University. Bailey wrote that

> Professor Menzel totally ignores the impressive testimony to the worth of Dr. Velikovsky's predictions contained in the recent letter of that outstanding scientist Professor H. H. Hess of Princeton.

Motz wrote that Velikovsky's

> writings should be carefully studied and analyzed because they are the product of an extraordinary and brilliant mind, and are based upon some of the most concentrated and penetrating scholarship of our period. (Juergens, 1966, p. 56-57).

Even with the support and recognition from the established scientists, political situations continued to bar his work from scientific journals. H. H. Hess, who had been president of the American Geological Society, recommended one of my father's papers for publication in *Proceedings of the American Philosophical Society*, the same journal that refused to print my father's rebuttal to Payne-Gaposchkin's misquotations of *Worlds in Collision* in 1952. The paper was held for six months by the Society and then returned to Hess saying they would not publish it. Despite the growing empirical evidence from astronomy, geology, archaeology and other fields, for the most part, the mainstream scientists continued using various unscientific and unethical methods of suppressing my father's work. They ignored his work, boasted about not having read the book, refused to make public his rebuttals of his critics' attacks, made statements completely unsubstantiated by facts, made false accusations and misquoted him. And, the libel continued. In 1969, Harlow Shapley was still writing letters calling my father a charlatan. Horace Kallen wrote about this in »Shapley, Velikovsky and the Scientific Spirit« in *Velikovsky Reconsidered*:

> Shapley's recent comments on Velikovsky, false on their face, seem to me variations of a persistent libel begun over twenty years ago, practically with the libeler's first contact with Velikovsky. It happens that I had a part in furthering the contact, and I cannot help feeling chagrin and disgust over its unbelievable consequences. ... As between Shapley and Velikovsky, the record for integrity is entirely in favor of Velikovsky. (p. 20, 32)

First, my father was excluded from professional meetings where his work was discussed. Then, he was invited to participate in symposiums where he was "set up" to be attacked and ridiculed by a biased panel who had no intention of giving him a fair hearing. Then, the scientists urged journalists and editors not to quote Velikovsky in articles because he is not a scientist and hence, not competent to judge scientific theories, and "editors are not inclined to reject such advice." (Juergens, 1966, p. 73).

In 1974, the American Association for the Advancement of Science (AAAS) held a symposium on "Velikovsky's Challenge to Science" in which my father participated. Having called him from the airport from where I shipped material to him, he implored me to come to San Francisco, anticipating vindication at the AAAS meeting. I joined my husband Sid and Carmel and we headed to San Francisco. The audience was receptive, the panel corrosive. This time Carl Sagan led the attack. One by one, he and other members of the scientific establishment delivered papers attempting to prove that my father's theories were theoretically impossible. The scientists at this meeting barraged my father with criticism, ridicule and sarcasm, but were unable to refute my father's theories. In fact, Sagan was even criticized for his inaccurate calculation of the odds against Velikovsky's cosmic catastrophes.

My father only belatedly understood that what he thought was to be a study of his works, was in fact, a gathering intended to once and for all put an end to the Velikovsky controversy, pronouncing him wrong!

During his lecture my father excused himself and briefly sat in the audience. That was the only time I had ever seen him rest in the middle of a lecture. He uncharacteristically read a paper which he had to provide the press with an advance copy. He received a standing ovation – but the press was biased. Concerned with his family, my father held up a packed press conference (prior to the debate) for over a half hour until our arrival from the east coast. He wanted us to enjoy and share what he thought would be a triumph. He used the time poorly at the press conference. My contacting Sullivan of the *New York Times*, and writing other letters, did nothing to change the bias, and in the face of an avalanche of attacks in the press, I felt my father's pain.

Carl Sagan left early to appear on the Johnny Carson Show and missed the debate with my father. After Carl Sagan talked, I walked over to him and said: "Anyone who bases his reputation on the destruction of Velikovsky is suicidal." He once told my father he did not want my father analyzing him, but to me he said nothing.

In December 1987, Sagan wrote to me, "I'm sorry to say that if anything the arguments in *Worlds in Collision* are even more untenable today than when we met."

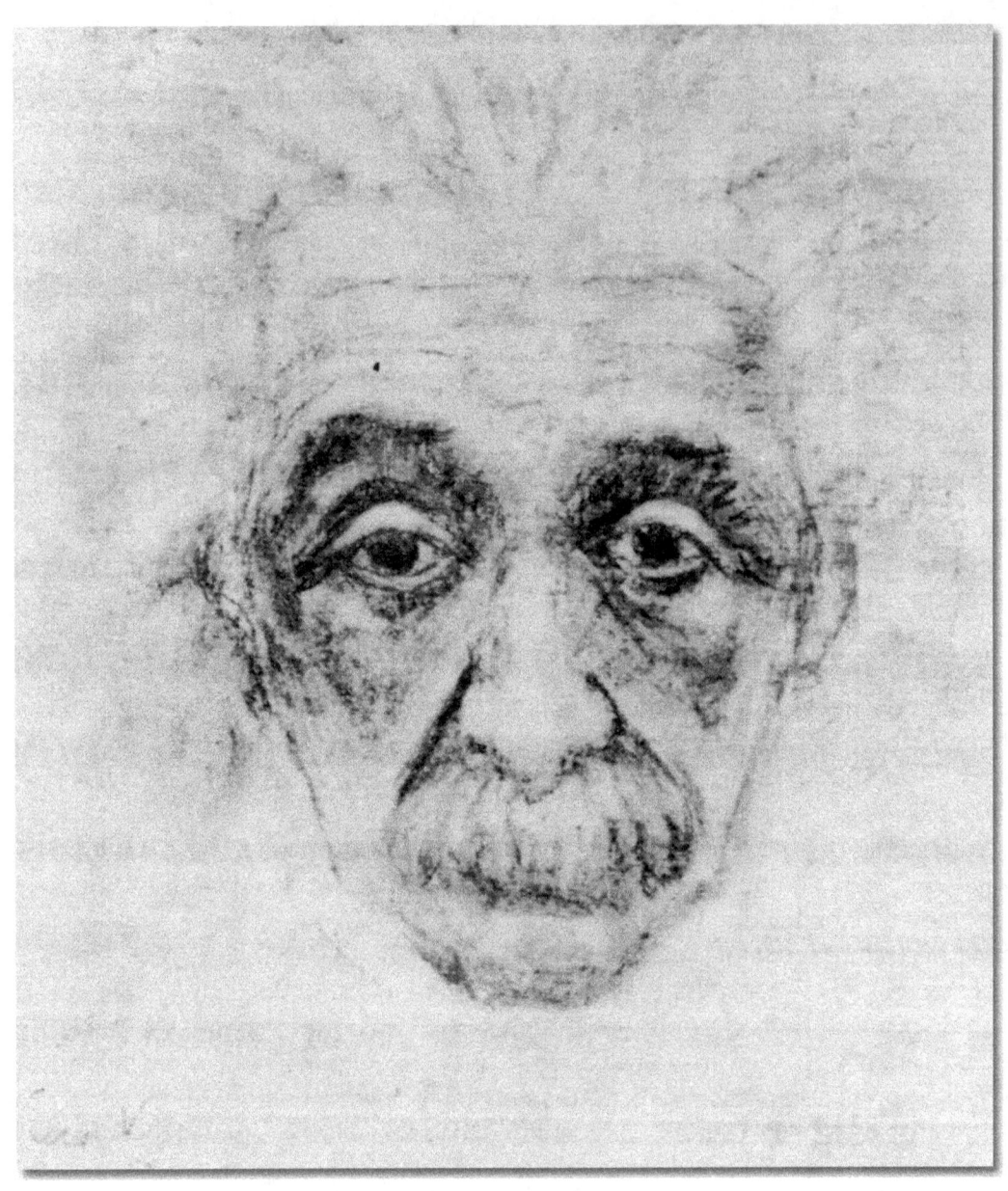

Albert Einstein
by Elisheva Velikovsky
Spring 1955

They also refused to print my father's rebuttal as promised, and it was excluded from the publication that documented that symposium, *Scientists Confront Velikovsky* (1977). His answers to them were published later, however, in two books, *Velikovsky and Establishment Science* (1977) and *Scientists Confront Scientists Who Confront Velikovsky* (1978).

It is interesting that while Carl Sagan repeatedly criticized my father's theories about cosmic collisions, one of his most recent articles is titled, »Long-Range Consequences of Interplanetary Collisions« (in *Issues in Science and Technology*, Summer, 1994, p. 67-72). Some of Sagan's sentences are curiously similar to *Worlds in Collision*:

> Comets have been associated with catastrophes in almost all cultures and since remotest antiquity. ... Collisions with the larger members of this population are catastrophic. The greatest danger is from impacts energetic enough to inject so much pulverized soil and rock into the stratosphere as to darken and cool most of the Earth, regardless of the impact location. (Sagan, 1994, p.67)

Despite the abuse and the silence on the part of the mainstream scientists and science editors, my father's theories are embraced by a growing number of scholars in every country around the world. His books became required reading in many university courses, especially those centering on interdisciplinary studies, study groups have been formed, and a growing number of scientists around the world continue to conduct research and publish findings that support Velikovsky's hypotheses.

Having moved to Princeton, one day in 1952, while walking along Carnegie Lake, my parents met Albert Einstein again. At first Einstein was reluctant to get involved with my father, just saying: "Ah, you are the man who brought the planets into disorder," but agreed to correspond with him by mail. In these letters my father and Einstein debated their positions on electromagnetism, catastrophism, Venus, the Earth's rotation, etc. While their opinions differed greatly at first, as time passed, Einstein's attitude gradually changed. Although he disagreed with my father on many issues, on others the area of disagreement diminished

My father wrote that, in one of his letters, Einstein referred to *Worlds in Collision* as a book of "immeasurable importance" that scientists should read. He also wrote that my father's research demonstrated that catastrophes could have been the result of extraterrestrial events, yet he doubted that Venus played a role in these events.

Like Lloyd Motz, and Harry Hess, Einstein could not be drawn into scientists' organized public distortion and abuse of my father or his work. On November 8, 1953, Einstein invited my parents to his home. This was the beginning of a friendship between them that lasted until Einstein's death in 1955. From the very beginning, he was interested in my father's ideas. And while he disagreed with my father about electromagnetism and the planets, he kept an open mind. Often, he would ask my father not to leave, but to stay longer and discuss more about celestial mechanics.

He and my father spent many evenings together debating various issues in astrophysics. My father even wrote an unpublished book about his relationship with Einstein, titled *Before the Day Breaks*.

Einstein read *Worlds in Collision* three times, and he read *Earth in Upheaval* and Shapley and others. He told him that while Shapley's behavior could be explained, it could not be excused. He also said to my father on more than one occasion that he believed the story of the suppression of *Worlds in Collision* must be made public. When my father began writing *Stargazers and Gravediggers*, Einstein reviewed and made notes on the manuscript. When he read my father's words that "The real cause of indignation against my theory of global catastrophes is the implication that celestial bodies may be charged," he (Einstein) wrote in the margin: "Ja" ("Yes") (*Stargazers and Gravediggers*, File III, Evenings With Einstein).

Letter 17 part 1

Letter 17 (transl.)

> 22. V. 54.
>
> Dear Mr. Velikovsky!
>
> Remarks on the part of your manuscript "poles displaced."
> The first impression is that the generations of scholars have a "bad memory." Scientists generally have little historical sense, so that each single generation knows little of the struggles and inner difficulties of the former generation. Thus it happens that many ideas at different times are repeatedly conceived anew, without the initiator knowing that these subjects had been considered already before. In this sense I find your patience in examining the literature quite enlightening and valuable; it deserves the attentive consideration of researchers who according to their natural mentality live so much in the present that they are inclined to think of every idea that occurs to them, or their group, as new. The idea of a possible displacement of the poles as an explanation of the change of climate in any one point of the earth's crust is a beautiful example. Even the idea of the possibility of a sliding of the rigid crust in relation to the plastic, or fluid deeper strata of the earth, was already considered by Lord Kelvin (and however rejected).
> The interpretation of the vote mentioned on pp. 159-160 as an attempt at a dogmatic fixation of the "truth" is not obvious to me. It is simply interesting for the participants of a congress to see how opinions concerning an interesting question are divided among those present. I don't think that the underlying idea was that the outcome of the voting would somehow insure the objective correctness of the outcome of the vote.
> From p. 182 on starts a wild robbers' story (up to p. 189) which seems to rely more on a strong temperament than on organized considerations. Referring to p. 191: Blacket's idea is untenable from a theoretical point of view. The remark about the strength of magnetization seems to be unjustified (p. 192); it could for example depend essentially upon the speed of cooling as well as on particle shape and size. The direction of the magnetic field during solidification must however quite certainly determine the direction of magnetization. Bottom 192 etc.: wild fantasy! from here on marginal remarks with pencil in the manuscript.
> The proof of "sudden" changes (p. 223 to the end) is quite convincing and meritorious. If you had done nothing else but to gather and present in a clear way this mass of evidence, you would have already a considerable merit. Unfortunately, this valuable accomplishment is impaired by the addition of a physical-astronomical theory to which every expert will react with a smile or with anger – according to his temperament; he notices that you know these things only from hearsay – and do not understand them in the real sense, also things that are elementary to him. He can easily come to the opinion that you yourself don't believe it, and that you want only to mislead the public. I myself had originally thought that it could be so.
> This can <u>explain</u> Shapley's behavior, but in no case <u>excuse</u> it. This is the intolerance and arrogance together with brutality which one often finds in successful people, but especially in successful Americans. The offence against truthfulness, to which you rightly called my attention, is generally human, and in my eyes, less important. One

Letter 17 (cont'd)

must however give him credit that in the political arena he conducted himself courageously and independently, and just about carried his hide to the marketplace. Therefore it is more or less justified if we spread the mantle of Jewish charity over him, difficult as it may be.

To the point, I can say in short: catastrophes yes, Venus no. Now I ask you: what do you mean when you request of me to do my duty in this case? It is not clear to me. Be quite frank and open towards me, this can only be good in every respect.

With cordial greetings to both of you,

Your

A. Einstein.

Letter 17
part 2

C o p y

September 17, 1954

Dear Professor Einstein:

May I renew our discussion?

At our last long conversation on July 21, you have acceded that the cause of the global catastrophes of the past could have been extra-terrestrial. You have found the behavior of Lexell's comet almost unbelievable.

The next step in my strategy is to show that the comets do not revolve as neutral bodies around a neutral sun. I quote from H. Spencer Jones:

"The presence of bright lines in the spectra [of comets] can only be due to a self-luminous body.... The electrical phenomena obtained by discharge through a Geissler's vacuum tube enable the assertion to be made with a high degree of probability that comet's self-luminosity is due not to an actual combustion, but to an electrical phenomenon."

More facts point to a charged state of the comets. The envelope (coma) of a comet contracts with the approach to the sun and expands with recession, though in the heat of the sun the reverse would be expected.

"There is good evidence that all particles in the comet influence the motion of each other. The configuration of the streamers in the tails... strongly indicate a mutual repulsion." (N. Bobrovnikoff, Comets, in Astrophysics, ed. Hynek, 1951, p.328).

Letter by Immanuel Velikovsky,
with Einstein's handwritten notes in the margin

12.III.55.

Lieber Herr und liebe Frau Velikovsky!

Sie haben mich bei Gelegenheit dieses unseligen Geburtstags aufs Neue beschenkt mit Früchten einer geradezu eruptiven Produktivität. Ich freue mich auf die Lektüre des historischen Werkes, das ja die Hühneraugen meiner Gilde nicht in Gefahr bringt. Wie es mit den Hühneraugen der anderen Fakultät steht, weiss ich noch nicht. Ich denke an das rührende Gebet: Heiliger St. Florian, verschon' mein Haus, zünd' andre an! Den ersten Band der Memoiren zu "Worlds in collision" habe ich bereits aufmerksam gelesen und mit einigen leider zu radierenden Randbemerkungen versehen. Ich bewundere Ihr dramatisches Talent und auch die Kunst und Gerissenheit von Thackrey, der den brüllenden astronomischen Löwen dazu gebracht hat, endgemessen den königlichen Schwanz einzuziehen unter nicht völliger Respektierung der Wahrheit. Ich würde glücklich sein, wenn auch Sie die ganze Episode von der drolligen Seite geniessen könnten.

Unvorstellbare Korrespondenz-Schulden und ungelesen zugesandte Manuskripte zwingen mich zu kurzer Fassung. Dank Euch beiden und freundliche Wünsche

Ihr
A. Einstein.

Letter 18

Letter 18 (transl.)

27. III. 55.

Dear Mr. and dear Mrs. Velikovsky!

At the occasion of this unpropitious birthday you have presented me once more with the fruits of an almost eruptive productivity. I look forward with pleasure to reading the historical book that does not bring into danger the toes of my guild. How it stands with the toes of the other faculty, I do not know as yet. I think of the touching prayer: "Holy St. Florian, spare my house, put fire to others!"

I have already carefully read the first volume of the memoirs to "Worlds in Collision," and have supplied it with a few marginal notes in pencil that can be easily erased. I admire your dramatic talent and also the art and straightforwardness of Thackrey who has compelled the roaring astronomical lion to pull in his royal tail somewhat, without completely respecting the truth. I would be happy if you, too, could enjoy the whole episode from its funny side.

Unimaginable letter debts and unread manuscripts that were sent in force me to be brief. Thanks to both of you and friendly wishes,

Your
A. Einstein

Wenn die Trudel aus dem Haus
Kennt sich Teufel nicht mehr aus
Ich, zur Ordnung nicht geboren
Fühl mich gänzlich, wie verloren.

Briefe regnet's ohne Zahl
Wer erbarmt sich meiner Qual?
Und der Postmann lacht mich aus
Denkt's ist ein verrücktes Haus.

G'schieht ihm recht, 's ist Gottes Walten
Warum kann er 's Maul nicht halten?
Darum geht's so toll bei der zu
Mich lässt alle Welt in Ruh.

Wie ich nun schon ganz verhornt
Hat die Ihne sich erbarmt
Kurzschrift und Orthographie
Klappern auch - ich könnt' es nie.

Alles kann sie, die Canaille
Sie verdient die Goldmedaille
Macht mein Schifflein wieder flott
Wenn's ihm geht, so lohnt ihr's Gott.

A. Einstein. 15. VI 53.

Letter 19

The Battle

Letter 19

> Wenn die Dukas aus dem Haus
> Kennt kein Teufel sich mehr aus
> Ich, zur Ordnung nicht geboren
> Fühl mich grauslich, wie verloren.
>
> Briefe regnet's ohne Zahl
> Wer erbarmt sich meiner Qual?
> Und der Postmann lacht mich aus
> Denkt 's ist ein verrücktes Haus.
>
> G'schieht ihm recht, 's ist Gottes Walten
> Warum kann er's Maul nicht halten?
> Darum geht's so toll hier zu
> <u>Mich</u> lässt alle Welt in Ruh.
>
> Wie ich nun schon gnung verharmt
> Hat die Göre sich erbarmt
> Kurzschrift und Orthographie
> Klappen auch – ich könnt' es nie.
>
> Alles kann sie, die Canalje
> Sie verdient die Goldmedaille
> Macht mein Schifflein wieder flott
> Wenn's ihn gibt, so lohnt ihr's Gott.
>
> A. Einstein, 15. VI 53

Letter 19 (transl.)

> When the Dukas is not at home
> No devil knows his stuff any more
> Me, not a born orderly person
> Feel terrible, quite lost.
>
> It's raining countless letters
> Who will have mercy with me?
> And the postman laughs at me
> He thinks the house is crazy.
>
> Serves him right, it's the workings of God
> Why can't he just shut up?
> This is why things are so wild here
> The world leaves <u>me</u> in peace.
>
> As I'm already careworn enough
> The gal takes pity
> Shorthand and orthography
> Work as well – I could never handle it.
>
> The canaille can do anything
> She deserves the gold medal
> Sets my little boat afloat again
> If he exists, do praise God.
>
> A. Einstein, 15. VI 53

In April, 1955, it was discovered that radio noises were being emitted from Jupiter, which my father had predicted years earlier. When he told Einstein of this discovery, Einstein offered to help him get radiocarbon tests of ancient relics to test his historical reconstructions.

A few days later Einstein died, but a letter was sent to the Metropolitan Museum of Art requesting these tests.

THE NEW YORK TIMES, MONDAY, JULY 21, 1969

Are Moon's Scars Only 3,000 Years Old?

By IMMANUEL VELIKOVSKY

Man, free from the bonds tying him to the rock of his birth, is about to make his first steps on the lunar landscape. It is an amazing achievement of man's technological genius, and with it the first stage of the Space Age (1957-1969) will be concluded.

These 12 years have been unkind to many accepted scientific theories of the solar system. Some of the most fundamental concepts are being summoned for revision.

In celestial mechanics, all new evidence has conjured against the concept — basic in science until very recently— that gravitation and inertia are the only forces in action in the celestial sphere.

The new discoveries are the interplanetary magnetic field centered on the sun and rotating with it; the solar plasma; the terrestrial magnetosphere that caused the moon to rock when entering and leaving the magnetic funnel; the enormously powerful magnetic envelope around Jupiter through which the Galilean satellites plow, themselves influencing the Jovian radio signals.

Who is the physicist that would insist that Jupiter, traveling with its powerful magnetosphere through the interplanetary magnetic field, is not affected by it? Or that the Jovian satellites are not influenced in their motions by the magnetic field of their primary?

And in cosmology the puzzling discoveries have been Venus's incandescent heat; its massive atmosphere (140 atmospheric pressures!); its retrograde rotation controlled by the earth (it turns the very same face to us when in inferior conjunctions), and its mountain-high ground tides (this is my understanding of the paradoxical altitude readings of the recent Venera 5 and 6), which also have caused it in the past to acquire a nearly circular orbit; Mars's moon-like surface and its apparent loss of a large part of its rotational momentum (Mariner 4), and the moon's active state—it is not a dead body cold to its core.

Establishing the Timing

All these discoveries unite to defend the thesis that the present order of the solar system is of recent date.

In divergence from accepted views, I maintain that less than 3,000 years ago the moon's surface was repeatedly molten and its surface bubbled. Since the nineteen-fifties, many unburst bubbles — domes — have been observed on the moon and gases have been found escaping from several orifices.

The moon has hundreds of hot spots and even its light is not all reflected solar light; researchers have come up with calculations that fluorescence would not account for the rest.

In thermoluminescence tests, it should be possible to establish the recentness of the last heating (melting) of the lunar surface. For that purpose, astronauts need to take samples from about three feet below the surface, to where the long lunar day hardly transmits any solar heat. Such tests could establish the time when the lunar surface was molten.

The moon has a very weak magnetic field; yet its rocks and lavas could conceivably be rich in remanent magnetism resulting from strong currents when in the embrace of exogenous magnetic fields.

Before their removal from the ground, the specimens should be marked as to their orientation in situ. Meteorites could not fall all similarly aligned. This simple performance of marking the orientation of samples, I was told, is not in the program of the first landing.

Despite the fact that there are no oceans on the moon and no marine life to give origin to petroleum hydrocarbons, I would not be surprised if bitumens (asphalt, tar or waxes) or carbides or carbonates are found in the composition of the rocks, although not necessarily in the first few samples.

Deposits of Petroleum

A visitor to the earth would not detect deposits of petroleum in the first few hours, either. I have claimed an extraterrestrial origin for some of the deposits of petroleum on earth; the moon did not escape the same shower. Only in a subsequent melting of the ground, such deposits would most probably convert into carbides or carbonates.

It is quite probable that chlorine, sulphur and iron in various compounds, possibly (the Deluge) and subsequently was covered for several centuries by water, which dissociated under the ultra-violet rays of the sun, with hydrogen escaping into space.

I maintain that—although not already at the first landing— an excessively strong radio-activity will be detected in localized areas, in those among the crater formations that resulted, I contend, from interplanetary discharges.

I also maintain that moonquakes must be so numerous that there is a bit of a chance that during their few hours on the moon the astronauts may experience a quake.

Some authorities (Harold Urey among them) claim that the scars on the face of the moon are older than four and a half billion years. The lunar landings will provide the answer: Was the face of the moon as we see it carved over four and a half billion years ago or, as I believe, less than 3,000 years ago?

If this unorthodox view is substantiated, it will bear greatly not only on many fields of science but also on the phenomenon of repression of racial memories, with all the implications as to man's irrational behavior.

IS NOT SURPRISED: Lord Shackleton, who comments on man's landing on moon.

Dr. Velikovsky is the author of "Worlds in Collision" and several other books arguing that cataclysmic astronomical events have helped shape human history. His books, despite their popularity, have evoked strong reactions from the scientific community.

THE NEW YORK TIMES, MONDAY, JULY 21, 1969

Technology a Spur to Changes in Religion

By EDWARD B. FISKE

(afternoon edition)

COMMENTS ON LANDING: Lord Shackleton, the Lord Privy Seal of Britain.

Copyright 1969 by The New York Times Company, Reprinted with permission.

In July, 1969, the *New York Times* prepared a special supplement for the moon landing and asked my father for a contribution. This contribution, an article entitled »Are Moon's Scars Only 3,000 Years Old?« appeard on the front page of the early edition of the July 21, 1969 *The New York Times*. By the afternoon/late edition, the article had been replaced by an article entitled »Technology a Spur to Changes in Religion« by Edward B. Fiske.

Before his death, my father received many invitations to speak, on college campuses around the country and abroad. He lectured at NASA's Ames Space Research Center in California in 1972, and in 1973, he spoke before scientists and engineers of the NASA Langley Space Research Center in Virginia. He was invited to speak before the 900 members of the Society of Harvard Engineers and Scientists at Harvard University, where twenty years earlier Harlow Shapley initiated his libelous attack. He was also awarded an honorary doctorate in arts and science from Lethbridge University in Alberta, Canada while attending a symposium. The papers delivered at this event and his acceptance speech were published in a book *Recollections of a Fallen Sky: Velikovksy and Cultural Amnesia* in 1978.

Honorary doctorate from Lethbridge University

Discoveries in many scholarly disciplines continue to confirm my father's advance claims, but today, geologists and astronomers are taking credit for "the new theories of catastrophism", the very theories that were first promulgated by my father in 1950. Right now, at the beginning of the 21st century, establishment scientists are still ridiculing and excluding him as they did half a century ago.

What follows are some examples of recent publications in the field of geology that illustrate this point.

In 1950, Velikovsky started a revolution that would affect how scientists view catastrophic events in the Earth's history. Yet, it has taken nearly half a century for geologists and paleontologists to recognize, as Velikovsky had in the 1940's and 1950's, that catastrophism is a reality. Consistent with what Velikovsky said in 1950, Stephen Jay Gould, in 1984, wrote in his article, »Toward a Vindication of Punctuational Change« (in *Catastrophes and Earth History: The New Uniformitarianism*, W. A. Berggren and John A. Van Couvering, Eds., Princeton University Press, 1984):

> gradualism has operated for the past one hundred and fifty years as a serious constraining bias in the history of geology. ... In defending the reality of discontinuous change, catastrophists, with their adherence to empirical literalism, had grasped an essential part of nature. (p.16)

Another scientist, D. J. Stevenson, has even remarked that "Catastrophism is now in fashion." (See Claude Albritton's *Catastrophic Episodes in Earth History*, Chapman and Hall, London, 1989, p. 173).

Just a glance at some of the hundreds of titles of both popular and academic publications would prove this to be true:

»Fire from the stars could spell global disaster ...« (John Gribbin, *New Scientist*, March 26, 1994, p. 16)

»Big Bang: NASA wants to fend off doomsday asteroids« (Tim Beardsley, *Scientific American*, November, 1991, p. 30)

»The Impact Giveth ... Did cosmic collisions help create life as well as destroying it?« (*Scientific American*, September, 1989, pp. 24-25)

»Geomagnetic reversal spurts and episodes of extraterrestrial catastrophism« (Pal Poorna & Kenneth Creer, *Nature*, Vol. 320, March 13, 1986, pp. 148-150)

The New Catastrophism (Derek Ager, 1993, Cambridge University Press)

Controversy – Catastrophism and Evolution: The Ongoing Debate (Trevor Palmer, 1999, Springer)

> *Impact!: The Threat of Comets and Asteroids* (Gerrit L. Verschuur, 1996, Oxford University Press)
>
> *Perilous Planet Earth: Catastrophes and Catastrophism through the Ages* (Trevor Palmer, 2003, Cambridge University Press)

These are just a handful of the hundreds of articles and books that deal with the topic of the "new catastrophism, also called neocatastrophism" and "liberated catastrophism". Some scientists are a bit leery of the association of the word "catastrophism" with the supernatural, and prefer the term "the new uniformitarianism": The more cautious types put the emphasis on the rarity of catastrophic events and call catastrophism "episodicity" or "periodicity".

Ager writes,

> Rudolf Trumpy, in his address as President of the International Union of Geological Sciences (1980) summed up the previous 10 years as follows: We have come a long way from the positivist, ruddy-faced uniformitarianism of ten years ago. Geologists are beginning to realize that even improbable events become probable during a sufficiently long time span. Catastrophism is probably the wrong word, but episodicity and periodicity ... loom large in the minds of today's geologists. (Ager, 1993, p. xvii).

Georges Cuvier, the father of 19th century catastrophic theory, along with other paleocatastrophists of the early part of that century, are now considered by some scientists to be the "precursors of today's neocatastrophists" (Albritton, 1989, p. 177). However, while Cuvier and his colleagues may have been the precursors of catastrophism, the pioneer of this new paradigm is Immanuel Velikovsky. Two decades before geologists and paleontologists even started to question the doctrine of uniformitarianism, Velikovsky had already researched, formulated, and published his interdisciplinary theory of the new catastrophism. Although many astronomers and geologists resist this truth, the emergence of a new paradigm was in large part triggered by Velikovsky's books *Worlds in Collision* (1950), *Ages in Chaos* (1952), and *Earth in Upheaval* (1955). He was the first scholar to draw to the attention of both the public and the scientific community the countless inconsistencies that appeared in the theory of uniformitarianism.

Eric Larrabee, in the introduction to *Stargazers and Gravediggers: Memoirs to Worlds in Collision* (Immanuel Velikovsky, William Morrow and Company, New York, 1982), writes that these anomalies

> have been receiving an amount of attention embarrassingly difficult to ignore. The anomalies are not Velikovsky's fault; they would have been there whether or not he had existed. He merely accelerated the process of their recognition by the decisive and thoroughly legitimate act of offering a new paradigm.

And, indeed, the accumulation of these anomalies skyrocketed as the inception of the space age approached a decade later. Establishment scientists were shocked to learn from the findings of the space probes that many of their assumptions about the solar system were wrong and that their gradualist theories were collapsing. By contrast, Velikovsky's predictions about the solar system were correct.

Claude Albritton writes that

> in the present ferment of ideas the principles of geohistory, biohistory and astrohistory are being subjected to critical re-examination toward the ambitious end of developing a unified theory for the unfolding of the universe (Albritton, 1989, p. 181).

A truly unified theory can only come about through merging the physical sciences, the social sciences, philosophy and history into one interdisciplinary science.

This is precisely what Velikovsky accomplished. He started this re-examination more than half a century ago and pioneered a new paradigm – a unified theory for the unfolding of the universe. William Mullen, in his chapter »The Center Holds« in *Velikovsky Reconsidered*, Editors of *Pensee*, Doubleday & Company, Inc., Garden City, New York, 1976, writes:

> Just as all the colors in a spectrum united make a white light, so all the disciplines in science united make one mode of knowledge. Not the least effect of a Velikovskian revolution should be to make scientists unable to forget that certain problems can be solved only if data from the most widely divergent fields are considered together. Interdisciplinary research will have to be regarded not as a luxury, but as an essential (p. 248)

The following evidence demonstrates that much of the geological and paleontological literature contains ideas and theories that were pioneered by Immanuel Velikovsky more than fifty years ago. While these theories are welcomed and embraced by scientists when their own colleagues make them, they are discredited and ignored when asserted by Velikovsky. Furthermore, while geologists and paleontologists freely admit their contempt for Velikovsky and his work, I will show that much of the evidence that supports their theories came into being because Velikovsky unearthed it, synthesized it and presented it into a useful paradigm.

In his book, *The New Catastrophism*, Derek Ager makes it very clear that he does not wish that his words be

> thought in any way to support the views of the 'creationists' (who I refuse to call 'scientific'). (Ager, 1993, p. xi)

He adds that

> Velikovsky seems to be on the side of the catastrophists, but I do not want to be associated in any way with such nonsense. This, together with the writings of the California 'creationists' are reason for my disclaimer at the beginning of this book. (p. 180)

The Battle

Against Ager's wishes, I nevertheless will quote in this section some of Ager's words, many of which are curiously similar to those written by Velikovsky many, many years ago. I do not suggest that Ager's work supports Velikovsky's theories, but rather, that Velikovsky's theories support Ager. In fact, if one did not know Ager's opinion of Velikovsky, one may actually believe, after reading his last book that Ager himself is a Velikovskian (or should I say 'was a Velikovskian'. Derek Ager passed away before his latest book was published in 1993).

Twenty years ago the British Association for the Advancement of Science met in Oxford, England on September 5th – 9th, 1988, the 80th anniversary of the devastation of Tungushka in Siberia in 1908 by a large meteorite. One of the highlights of the meeting was a session called "Catastrophes and Evolution". *The New Scientist*, in its September 8th issue of that year, reported on some of the papers presented at this session, particularly one presented by Dr. Victor Clube of the University of Oxford. *The New Scientist* reported that

> Recurrent 'cosmic winters', little ice ages occurring every thousand years or so as a result of bombardments from space, may be a feature of Earth's history, according to Victor Clube, of the University of Oxford. He says that stories of fire from heaven in myths, legends and historical records should be taken seriously. Rather than being exceptional, he says, catastrophes are normal on all time scales longer than a thousand years. ... Clube quoted historical sources from that time that talk of 'stars in the sky [that] were seen throughout the whole world to fall towards the Earth, crowded together and dense, like hail or snowflakes. A short while later, a fiery way appeared in the heavens; and after another short period half the sky turned the color of blood.' Six centuries before this, according to Clube, Europe and particularly Britain, experienced 'widespread destruction, followed by years of migration, darkened skies and a dark age'. Why was this cultural dark age so severe? Clube pointed to Chinese references to a 'strange comet' in 442, and to 'alarming' meteoric events in 524. 'By far the most curious fact is the extreme intensity of the historical reverse ... for it is generally agreed that the level of civilization previously enjoyed was not restored again for around 1300 years.' Did human civilization nearly go the way of the dinosaurs, and were falling stars to blame? (p. 34)

It is obvious from this passage that Clube's ideas mimic Velikovsky's thesis in *Worlds in Collision* (1950), which predated Clube's by nearly thirty years.

Velikovsky asserted,

> ... (1) that there were physical upheavals of a global character in historic times; (2) that these catastrophes were caused by extraterrestrial agents; and (3) that these agents can be identified. (Velikovsky, 1950, Preface)

Clube states that his theory is supported by a historical source that said "the sky turned the color of blood." The following quote demonstrates that this idea was stated first by Velikovsky more than half a century ago in *Worlds in Collision*:

> One of the first visible signs of this encounter was the reddening of the earth's surface by a fine dust of rusty pigment. In sea, lake, and river this pigment gave a bloody coloring to the water. Because of these particles of ferruginous or other soluble pigment, the world turned red. (Velikovsky, 1950, Part I, Ch. 2, »The Red World«)

The above quote also shows that Velikovsky even takes the idea a step further, offering an interpretation for this phenomenon, while Clube leaves the idea hanging. Similarly, Clube speaks of stars that fell to the earth like falling hail and snowflakes. Again his ideas bear a striking resemblance to Velikovsky's descriptions:

> Following the red dust, a 'small dust', like 'ashes of the furnace', fell 'in all the land of Egypt' (Exodus 9:8), and then a shower of meteorites flew toward the earth. Our planet entered deeper into the tail of the comet. The dust was a forerunner of the gravel. There fell 'a very grievous hail, such as has not been in Egypt since its foundations' (Exodus 9:18). Stones of 'barad', here translated 'hail', is, as in most places where mentioned in the Scriptures, the term for meteorites. (Velikovsky, 1950, Part I, Ch. 2, »The Hail of Stones«)

And, what about Clube's theory about widespread destruction, followed by years of migration, darkened skies and a dark age? This is what Velikovsky said in *Worlds in Collision*:

> If 'the Egyptian darkness' was caused by the earth's stasis or tilting of its axis, and was aggravated by a thin cinder dust from the comet, then the entire globe must have suffered from the effect of these two concurring phenomena; in either the eastern or the western parts of the world there must have been a very extended, gloomy day. ... Nations and tribes in many places of the globe, to the south, to the north, and to the west of Egypt, have old traditions about a cosmic catastrophe during which the sun did not shine; ... Tribes of the Sudan to the south of Egypt refer in their tales to a time when the night would not come to an end. (Velikovsky, 1950, Part I, Ch. 2, »The Darkness«)

The evidence speaks for itself. Velikovsky's theories are being advanced by geologists without any references or credit made to Velikovsky. While they pronounce Victor Clube a genius, they denounce Velikovsky a charlatan. An illustration of just this point is geologist, Derek Ager's comparison of the two in his 1993 book, *The New Catastrophism*. Ager claims to have read *Worlds in Collision*, and about Velikovsky he writes:

> The height of such idiocy is illustrated by the cult of Velikovsky, who postulated major collisions up into historic times. ... I will not encourage such pseudoscience by giving a reference. ... He does not present any geological data whatever to support his views. He merely cites very out-of-date authorities (some 18th century!) in a highly selective manner. ... (Ager, 1993, pp. 179-180)

First of all, it is important to note that Velikovsky does indeed cite both the geological and paleontological literature in his books. Just some of the geologists and paleontologists he cites are: R. A. Daly, F. K. Mather, C. R. Longwell, A. Knopf and R. F. Flint, D. F. Hertz, and of course, Cuvier, J. A. Deluc, G. F. Wright. Furthermore, he stated on the concluding pages of *Worlds in Collision*,

> Of these things I intend to write in another volume, where problems of geology and paleontology and the theory of evolution will be discussed. (Velikovsky, 1950, Epilogue)

In this way, Velikovsky referred the reader to his third book *Earth in Upheaval*. Ager would have discovered in this book a thorough discussion of geological data to support the notion of catastrophic change on our planet. In any event, right after denouncing Velikovsky, on the same page, Ager writes:

> We come then to <u>scientific</u> [my emphasis] views of extra-terrestrial impacts.

One of these scientific views is that of Victor Clube:

> Victor Clube, suggested (1978) that our galaxy had a violent history and later he and Bill Napier argued for a theory of terrestrial catastrophism which really started the recent interest in asteroid impacts on the Earth (Napier & Clube 1979). Later they produced a remarkable book (1982) which was subtitled *A Catastrophist View of Earth History* presenting a convincing argument for a giant comet that terrorized mankind in prehistoric times. They put together a great deal of evidence from a variety of sources that speaks of a violent past and a hazardous future. (Ager, 1993, p. 180)

Victor Clube's work, Ager thought, was convincing, while Velikovsky's was nonsense. Ager claims Velikovsky did not cite the geological literature, but Velikovsky did. This leads us to conclude one of two things: either Ager, like other establishment scientists, did not read *Worlds in Collision*, or he and others have taken credit for Velikovksy's theories.

The most popular and most frequent method of discrediting Velikovsky is to discredit his sources of data. From the very beginning he was criticized for quoting outdated, unscientific material, especially the Old Testament. Harlow Shapley, the Harvard astronomer who relentlessly attacked Velikovsky, accused him of basing his conclusions on incompetent data. At present, though, geologists, paleontologists and other scientists freely quote biblical passages to strengthen their own catastrophic theories, yet they continue to criticize Velikovsky for the same.

For example, in the quote above, Ager attacks Velikovsky for quoting outdated sources ("some 18th century!"). Why are 18th century observations more out-of-date than 19th century observations? Ager certainly makes numerous references to the observations of

Georges Cuvier published very early in the 19th century. Cuvier published his *Essay on the Theory of the Earth* in 1827, and five years earlier, in 1822, he coauthored a book with A. Brongniart. Perhaps Ager and other geologists forget that

> ... we have to keep in mind that our ancestors' eyes were as good as ours. (R. Hooykas »Closing Remarks« in *Contributions to the History of Geological Mapping*, INHIGEO Sympos., Akad. Kiado, Budapest, 1984, pp. 423-29)

Speaking of outdated references, it is also interesting to point out that Ager, like Velikovsky, often cites biblical references to support his own catastrophic theories. He says:

> It is not my usual custom to quote holy writ in my geology, but the text referred to is highly relevant in this chapter. (Ager, 1993, p. 79)

Ager then cites biblical scripture, and even offers a Velikovskian interpretation:

> ... it is not my business here to discuss the many contradictions in that strange book. Nevertheless, it may be said that disastrous tsunamis or storm surges are commonly preceded by a withdrawal of the sea, which might have been long enough to allow the Israelites to make their escape from Egypt if they were very quick. (p. 79)

Further along in his book, Ager links catastrophism with incidents recorded in the Bible when he writes:

> The record of earthquakes in historic times is also interesting. Ambraseys (personal communication 1968) deduced from ancient writings and from whether or not ancient buildings were designed to resist them, that earthquake frequency since the coming of city dwellers has been distinctly episodic. Though Joshua attributed the fall of the walls of Jericho to the blowing of trumpets and the shouting of the Israelites (before they slaughtered every man, woman, child and animal in the city in the charming way of the Old Testament), it seems more likely that an earthquake was responsible. In fact further earthquakes shook Jericho in the years AD 600, 1033 and lastly on 11 July 1927. A similar cause may be blamed for the fall of Sodom and Gomorrah (p. 168).

Ager also supports another scientist who discusses ancient knowledge about evolution when he writes:

> it has recently been pointed out by Alan Batten (in *New Scientist*) that since living 'serpents' have rudimentary legs, God's verdict on the serpent in Genesis that 'upon thy belly shalt thou go' indicates an early recognition of evolution." (Ager, 1993, p. 9)

The Battle

In light of the above references to Ager, the quote below illustrates the contradictions throughout his book with respect to the relevance of biblical references. It seems that Ager is genuinely conflicted over the role that biblical references play in his "new catastrophism". In the Introduction of the same book he writes:

> This is not the old-fashioned catastrophism of Noah's flood and huge conflagrations. I do not think the bible-oriented fundamentalists are worth honoring with an answer to their nonsense. No scientist could be content with one very ancient reference of doubtful authorship. For different reasons I do not go along completely with the modern fashion of cosmic collisions or with dark stars causing periodic mass extinctions. To me The New Catastrophism is mainly a matter of periodic rare events causing local disasters. (Ager, 1993, p. xix)

Perhaps one reason why geologists and paleontologists have resisted the use of historical documents in interpreting geological and paleontological findings is the long-standing approach to teaching this field. Geology students learn that to understand the past, one must study the present. They write the earth's history by examining not the past, but the present. During the past decade, this scientific method of geology has been criticized by many of their own, even Derek Ager, when he wrote:

> One must constantly ask oneself 'Is the present a long enough key to unlock the secrets of the past ?!' (Ager, 1993, p. xviii)

It is remarkable that more than half a century ago Velikovsky summed up one of the major problems in the field of geology that took the geologists themselves more than thirty years to acknowledge, that they have to examine the past to understand the present. Velikovsky wrote in *Worlds in Collision*:

> Traditions about upheavals and catastrophes, found among all peoples, are generally discredited because of the shortsighted belief that no forces could have shaped the world in the past that are not at work also at the present time, a belief that is the very foundation of modern geology and of the theory of evolution. (Velikovsky, 1950, Part II, Ch. 6, »Of "Pre-existing Ideas" In the Souls of Peoples«)

To support his observation, Velikovsky cites H. F. Osborn, who wrote *The Origin and Evolution of Life* in 1918:

> "Present continuity implies the improbability of past catastrophism and violence of change, either in the lifeless or in the living world; moreover, we seek to interpret the changes and laws of past time through those which we observe at the present time. This was Darwin's secret, learned from Lyell." (Velikovsky, 1950, Part II, Ch. 6, »Of "Pre-existing Ideas" In the Souls of Peoples«)

Velikovsky correctly asserts:

> It has been shown in this book, however, that forces which at present do not act on the earth, did so act in historical times, and that these forces are of a purely physical character. Scientific principles do not warrant maintaining that a force which does not act now, could not have acted previously. Or must we be in permanent collision with the planets and comets in order to believe in such catastrophes? (Velikovsky, 1950, Part II, Ch. 6, »Of "Pre-existing Ideas" In the Souls of Peoples«)

After more than 50 years, perhaps geologists are beginning to understand what Velikovsky tried to tell them years ago when he wrote:

> If, occasionally, historical evidence does not square with formulated laws, it should be remembered that a law is but a deduction from experience and experiment, and therefore laws must conform with historical facts, not facts with laws. (Velikovsky, 1950, Preface)

Consistent with Velikovsky's opinion, John Van Couvering of the American Museum of Natural History, in his book *Catastrophes and Earth History: The New Uniformitarianism*, (edited with W. A. Berggren, Princeton University Press, Princeton, N. J., 1984), urges geologists to recognize that

> the past may be the key to the present

and that

> gradualists must accept that human experience now includes the discovery in the rock record of unpredicted events, paleomagnetic reversals being salutary examples. (p. 5)

Velikovsky played a major role, whether mainstream scientists want to admit it or not, in forcing geologists to rewrite the Earth's history. Einstein told Velikovsky that

> the scientists make a grave mistake in not studying your book (*Worlds in Collision*) because of the exceedingly important material it contains." (*Stargazers and Gravediggers*, File III, »Evenings With Einstein«)

Yet it is also a disgrace that for years, scientists ridiculed Velikovsky or systematically ignored his work in a conspiracy of silence, while taking credit for his theories.

Depression

Studio at 558 West 113th Street

Sigi

Aba never attacked his enemies, no matter what tactics they resorted to to undermine his works and character. Accused of doing "more damage than prostitution and communism combined," paid ads for his books were refused in scientific journals. He could not cope with the derision and accusation that he is a charlatan, while having no way of getting scientists to read his books. Honest, educated, and capable of crossing boundaries and opening new vistas of research, to be accused of dishonesty was beyond his ability to cope with, and he sank into depression.

My father's first bout with depression occurred when he returned from Israel shortly before the publication of *Worlds in Collision*, after Shapley's "reversal" toward my father. Having been mistreated by a dishonest man involved with his Tel-Aviv property, my father unable to tolerate being deceived and humiliated – feelings he had difficulty coming to terms with – began to emotionally regress. Although he himself was a psychiatrist he refused help and sank.

His first depression pulled my mother right along. Arriving from Israel with a bad case of uncontrolled diarrhea Aba lost weight. Every doctor we took my father to, he insisted my mother was the sick one and sent her into the doctor's office. She was remorseful for not having remained in Israel and putting my father in a sanatorium there, and sank deeper and deeper into despondency.

Sid and I moved in with my parents into their second floor apartment on 558 West 113th Street. We slept on a mattress placed on the floor next to my mother's bed. My mother's regression was spiraling out of control at which point Paul Federn, an early associate of my father, a stocky man with a cane, was contacted and came to visit. He arranged for my mother's hospitalization in a private sanatorium in upstate New York. There, she had hallucinations about Nazis spying on her until shock treatments took effect. She had been forced to suffer the indignities of a straight jacket, and later, a small enclosed yard. She was suffering until, no longer stopped by my father who had insisted she'd go blind from shock treatments, I gave permission for the treatments and her suffering ended. Her improvement was so dramatic that we too were shocked. I had given her a piece of paper the previous week with the word "police" scribbled on it in Hebrew to allay her fears. When she pulled it out of her pocket after six treatments, it was meaningless to her. She once again admired the fall colors.

While my mother was hospitalized, my father made several trips to the hospital on Long Island, each time returning home. When he finally agreed to sign himself in, my uncle Sigi, who had been so helpful when we first arrived in the United States, accompanied him pronouncing, "Now you are where you belong." He was not soon forgiven for this remark.

Recovered, my parents were reunited – my mother making up her mind never to join my father on that type of trip again. Although my father suffered six more bouts of serious depression, it was not until after his death, when my mother suffered the stroke, that she again became depressed.

After dealing helplessly with his first depression, I shied away from my father during the next depression. While he was hospitalized, my mother stayed with us. She talked incessantly about my father. My analyst at the time recommended that I send my mother to her own home, for which she did not forgive me until my second analyst helped me help my father. While I remained uninvolved with his depression my father was hospitalized in an upstate New York hospital. While there he swallowed an excessive amount of medication. His stomach pumped, my mother was informed. While in the hospital the psychiatrist in charge took advantage of my father's state of mind and persuaded him to give him several of Freud's handwritten letters.

During the next depression, while at home, my father swallowed a bottle of pills. My mother, unable to arouse him, found a note he had written in small irregular handwriting and the empty bottle. She summoned an ambulance. He was hospitalized at Princeton Hospital and kept there for his "diabetes."

Aba suffered seven depressions between periods of productivity and optimism. Told that the depressions could not be avoided and would become more frequent, it was not until the help of the psychiatrist who trained me that the process was reversed. Both my parents had weekly phone conversations with the psychoanalyst. Having been accustomed to a man of distinction in her husband, this doctor psychologically filled the spot for my mother.

Before each hospitalization of my father, there were months of his slowly going downhill, a process the doctors could not stop. Eventually, we came to recognize the signs of an oncoming depression. Indecision was a sign that he was depressed for he was decisive and opinionated when stable. When the depression was beginning, he fidgeted with something in his hand, belying tension. If, at that stage, my mother would call attention to his procrastination, he would sink further. It was not possible (and not good for her) to keep her from repeatedly expressing her feelings. My father's facial grimaces were warning signs that the depression was deepening, and my spirits sank.

As my father's depressions worsened, he stopped eating, refused to take his medications, became physically agitated, thin and howling.

When he finally came out of the hospital, my parents stayed with us for several weeks making occasional daytime trips to their home, until the depression was lifted and they were ready to move back to Hartley.

In the last years of his life when Aba would slide into depression, I would be by his side insisting that he walk with me to keep his diabetes under control and to help his depression. Pulling him by the coat collar with great effort out of the house onto the porch and onto the street, we walked in all kinds of weather, as he talked to me of his thoughts, fears, concerns and recollections. I loved my father.

During the last depression, Aba suffered within the mire of his mind. My mother said the worst day of her life was when my father during a bout with depression stopped at a young lawyer's office and blurted out all his fears.

He waited for me every day until I returned from my New York office. In the hospital he was certain I would not return, having met with a terrible end at the hands of the enemy. He felt accused of having stolen another patient's dentures for his own use, and insisted I leave my purse open when departing, proving to the staff that I was not acting on his behalf and absconding with the dentures. I brought his books to him to inscribe on my son's birthday. After inscribing one "from much maligned, Immanuel", I took the rest home uninscribed.

My mother relied on me to deal with the doctors and visit him daily. I got permission to take him out for walks, which we did even during the frigid winter. My mother was particularly concerned with his reputation and did not want his name on the door of his room. After her initial visit, when she saw him in a white hospital gown among the deranged, she did not return until the day he was discharged. Before his discharge, however, although my father had rallied, his doctor recommended shock treatment – a punishment worse than death for a man whose brain was so ladened with information. I protested and took him home.

I was living in close physical and emotional proximity to my parents in New York until 1949 and in Princeton since 1955. My sister lived in Israel, coming to this country on visits. My husband and three young children were affected by neglect caused by my constant distraction and total absorption by the trauma of the incredible household at Hartley. My mother wrote pages complimenting my help to my father, which she considered wonderful.

When my father was not depressed things were exciting and positive and much activity rushed around him and his work. The rest of the time, it was chaos.

Together my parents soared in excitement during good times and together they sank in despair when my father's self-esteem was mutilated.

If something happens to me *please*, _never_ feel in any way guilty. I love you and understand how much you suffer. You will be soon well and strong again and even write your books and see that the best is yet to ~~she~~ be even without me, but hopefully with me together

Shenik forever!

December 72

Better Times

Aba at Hartley

Photo by Noveck 1975

In happier days, my father's fame led to some pleasant anecdotes.

While shopping at a shoe store in the Princeton Shopping Center, the salesman whispered to me: "See that man over there, he is another Einstein." It was my father shopping for shoes for my sister. He regularly bought my sister shoes and mailed them to her in Israel. I was happy to unexpectedly run into my parents in a store or on a Princeton street. They always greeted me with delight. On one occasion cub scout den parents met in our living room, and some time later, when my father was going to lecture on the Princeton University campus, I received a call from one of the den mothers informing me that Velikovsky was going to lecture on campus. She added that since she had seen Velikovsky's volumes on my living room shelf she thought I'd be interested in meeting him. My daughter, Carmel, upon telling someone in Philadelphia that she lived in Princeton, was told, "give my regards to Velikovsky." The person did not suspect that she knew him, much less that he was her grandfather.

When well, Aba was optimistic behaving like a millionaire. He selected for us our Deerpath house in which we've lived since 1955. When *Worlds in Collision* was doing well, he shared his fortune with his family, including my in-laws, sending my husband's father money to start a business. He spent little on himself, except once. Driving past a car showroom, he saw a Buick he liked. That day he drove it home to show my mother. It was the car he kept to the end of his life.

One day, he bought me a new car. He had gone to purchase a car for my mother and, while there, impulsively bought one for me – a bright red car, inside and out. He also selected a car to ship to my sister in Israel. It was yellow, but Shulamit had just bought a car. My mother, I suspect, did not feel special because whenever he gave her a present he'd also think of the children and grandchildren. When he bought my mother a gold watch, he also bought the same watch for me, for my niece-in-law, my niece, and for the young woman who worked for him at that time. Having a conservative nature financially my mother must have found it difficult to see my father duplicating every present several times.

A man of the highest integrity, Aba never cheated anyone. He liked an occasional bargain although he bought the Pelican Island house for the asking price and paid more than the asking price for the Hartley house. In the jewelry store where he bought the gold watches, he asked the storeowner to reduce the cost of a grandfather clock.

Aba was self-sacrificing and generous. He showed his generous spirit in other ways as well. Walter Federn, Paul's son, an egyptologist, was of great scholarly help to him. A stooped, anorexic, bushy-eyebrowed man, when his father, Paul Federn, was diagnosed as having a brain tumor, Paul asked Walter if he wanted to die with him. Walter said, "No!" Then, Paul, without telling my father what he was planning, asked Aba to take care of Walter should something happen to him. Aba agreed. The next day Paul shot and killed himself. From that point on, Aba looked after Walter until Walter's death.

Music room at Hartley
Ima's wood carving on window sill
Photo by Noveck 1975

One of the last family occasions I spent with my father, a few months before his death, was at a Met performance of Don Giovanni at the Lincoln Center. Tickets were scarce, hence one seat I purchased was in the first row directly behind the conductor. In his usual self-sacrificing style, my father opted to sit behind the conductor. There was no use arguing with him. He took the worst for himself. He told a story of life as a youngster. When food was served in his home, he and his two brothers fought over the smallest portion even though they all liked the food and wanted the larger portion.

My father liked to share his fortune, but when depressed, predicted the poor house for himself and his family. Writing over to me the financial rewards from the Spanish translations of his books, when he became depressed, he proclaimed me a wonderful daughter upon returning the rights to him.

My father believed in God. In time of personal indecision, he would open the Old Testament for symbolic answers. My mother did not write on the Sabbath, and neither of them conducted business transactions on the Sabbath. My mother once admonished me, "What would people think seeing Velikovsky's daughter at the bank on Saturday?"

My father had a disdain for "kitchen religion", as he called it, involving food and its preparation. My parents ate non-kosher chicken in restaurants although the house was kept kosher. At the end, my father particularly enjoyed steaks at a local restaurant. When in the 1930's in a European restaurant, my parents commented to each other in Hebrew that perhaps the food in front of them was horse meat. To their embarrassment, a man at the next table reassured them in Hebrew that it was not horse meat. After that, a Hebrew speaking person within earshot they referred to as "sus" (Hebrew for "horse").

My parents attended the Princeton Jewish Center on Yom Kippur and Rosh Hashannah. My father was often given the honor of holding the Torah, and when frail he opened and closed the ark. Sometimes my parents read prayers and fasted at home.

Later in life my father had to watch his diet carefully because of diabetes. The day before he died, while listening to chamber music, he ate a banana. Frequently, he said that if one likes to eat something, that meant it was good for that person. The banana was not good for him. Perhaps in retrospect, though, having feared a stroke or senility like his mother, and compared to the nightmare my mother experienced during the final months of her life, enjoying the banana, listening to chamber music, and then dying was not a bad end. But he died too young – his mind clear, his brilliance intact.

My parents drove on excursions on the Sabbath, but refrained when my sister was visiting them. When my father, whom I taught to drive in Princeton, took the wheel he relied on my mother to alert him of red lights and stop signs. He was known to speed and disregard stop signs. My mother, an over cautious driver, on the other hand, good humoredly commented that just once she wished she'd get a speeding ticket.

After her cataract surgery, my father, contrary to the doctor's opinion, did not permit her to drive because of her lacking peripheral vision. Eventually, we gave her one of our cars and titled it in her name, thus ending her imprisonment, as she called it. She often

The Pelican Island House   Aba's Buick

Aba on Pelican Island
Photos by Leroy Ellenberger

drove a few blocks, with great joy, to the Princeton Shopping Center, where she did not have to parallel park. My father eventually let her drive part way to their seashore house. After my father died, Jan, her personal assistant, drove her.

Aba was interested in acquiring real estate in the Princeton area but somehow let many opportunities elude him. Later, when driving in the area, he often expressed regret at not having bought land. My mother reassured him that he had bought the sky, which pleased him. He regretted often and once said, "I regret that I regret."

A few years before he died, my father purchased their house on Pelican Island claiming the ozone in the air was healthy and walking in that atmosphere was invigorating. I never liked the house and felt isolated from my life whenever there. One of the last occasions that I visited there, my daughter Carmel came with me. Aba became jealous of sharing my time with her, even for a few moments.

One summer day I spent at the seashore working together with my sister on gathering quotes in support of my father's theories. There was a lightning storm. It was close, loud and frightening. My father, who liked storms, came in to the room to comfort us. He was smiling – storms gave him an outlet for his strong feelings. I stayed overnight and heard through the thin walls my father sighing, "Oh! Mamitchka."

Aba could be quick to judge, at times without accuracy. At a very young age I sent a postcard to a children's radio program with my sister's name and teen birthdate on it. Aba accused me of malice. He could be a poor judge of character and when he impulsively liked someone I was cautious. He was frequently disappointed in his friends and said, more than once, that one should be weary of one's friends.

When he felt impressed or loving, it was easily detected on his face. When disapproving, he'd make no effort to mask his disdain. To the end of his life, his mood affected his family. When my father was in good spirits and in a creative mood he'd sit in the "history room" at Hartley – a room where the galleys to his historical volume, *Ages in Chaos*, remained on his desk untouched for ten years. Sitting in a red webbed chair in front of a small book-stand, Aba hummed as he worked on a manuscript with a sharpened pencil. The atmosphere was electrifying – his mood permeated the house. My mother was at her best and invited me to take a look. Feeling admired, my father continued to work, although his working span grew shorter as he got older.

My mother was destined to marry a great man, and if his reputation faltered, so would hers. She did not want to go down in history as the wife of what the scientific establishment called a "charlatan." She was devoted beyond measure, however, her punctual personality and her over-involvement with my father's work proved to be disastrous. My mother was often impatient. When she wanted something done there was an immediacy to it.

My father, on the other hand, behaved as though he were immortal. When he died he left a list he had written a few days earlier containing over sixty items in need of his attention – not a list of a man who thought he was about to die. He had a poor sense of time, catching trains the last minute, while my mother was punctual or ahead of time.

Aba, Jan and Lewis
Photo by Leroy Ellenberger

Ralph Juergens
June 1967

Aba and Lynn Rose

My mother was ambitious for my father and she often labeled him his own worst enemy. His procrastination, missing – purposely or negligently – deadlines for articles, corrected galleys, letters or whatever, caused my mother anguish. The galleys to *Ages in Chaos* were a familiar sight on his desk for over ten years. He alluded to not wanting to add the wrath of historians to that of physicists, geologists, etc. and as time passed the work and revision seemed almost insurmountable.

To the end of his life my father proudly said that he would not change one word in *Worlds in Collision*. He was confident that his theories, research and conclusions were correct although never claiming infallibility. Vindication mounted with space exploration – but he was pariah among scientists to the end. How could one man be right about so much? A real embarrassment ... It was fallout from one central idea.

To illustrate how incredible it is to disregard one wrong premise in a theory my father told a story of a man who complained to the baker that he found a fly in his raisin bread, to which the baker responded, "Bring back the fly and I'll give you a raisin."

Aba was not without supporters. Lewis Greenberg, a devoted friend, gave him consistent positive information surrounding the Velikovsky controversy, and my father liked talking to him on the telephone for hours at a time. Professor Greenberg was often frustrated with my father's procrastination – not receiving promised articles for *Kronos*, a pro-Velikovskian journal published at Glassboro State College.

Prof. Lynn Rose helped in the division of the archives between Jerusalem and Princeton.

Ralph Juergens, a devoted and capable man began compiling *The Test of Time*, my father's unpublished book about confirmations of my father's predictions. He died too young.

Aba attracted young scholars as well in the traditional mentor/disciple relationship. Devotees who wanted to work with him often lived in a set of rooms on the third floor of the Hartley House.

Jan Sammer, a resilient person, brilliant and devoted, was able to roll with the punches. When Aba procrastinated, Jan worked on his own. Without friends or clear identity, that period of his life was entirely spent at 78 Hartley. His parents, convinced that he had been snatched by a cult, pressured him to leave. A handsome, charming man he eventually developed friendships in Princeton and, to his parent's dismay, remained at Hartley.

Jan loved my father. During Aba's last bout with depression, he sat by his bedside to comfort him. Months before Aba died, Jan left for his home in Canada. When he returned to help my mother, remorseful at not having stayed to help my father, he was distracted with many interests and commitments and he did not sympathize with my mother's personality – resenting her impatience and expectations.

Other ardent followers gave up everything to work for my father. My mother became jealous of Kathy, a heavy, dark-haired, young woman who sat at the kitchen table with my father as my mother served food. My mother objected to cooking for anyone except

Aba inscribing book for Eddie Fisher

for my father. Aba, a faithful husband, read the daily mail with Kathy. My mother, who had been accustomed to sharing the mail with my father as well as often being on the phone extension (with everyone's knowledge and approval) felt replaced. Kathy left soon after.

Lorraine was a smart, helpful young woman who assisted Aba at the AAAS meeting – duplicating papers, etc. Her punctual temperament clashed with Aba's procrastinating nature. Whenever he proclaimed that they would begin working the next day at a particular time, it had nothing to do with what he actually did – although he meant it when he said it. She could not adjust, and in her youth, lacked understanding that my father's health was more important than his writings, and that his mind was encased in the body of an old man, depleted by depressions triggered by the ruthless attacks on his works and his character.

Richard Heinberg, a frail young man, pleasant in manner, had come to work for my father four days before my father died. During those four days, my father asked Richard – should something happen to my father, would Richard stay and help my mother? He agreed and stayed. My mother wanted to remain at Hartley. Richard lived in the attic where he conscientiously practiced the violin, much to my mother's chagrin, for she rarely found time to practice her violin. Occasionally, she played duets with Richard and when including him in her chamber group, she expected him to make up the work time in the evening. Daytime was reserved for working on my father's manuscripts, and my mother often sat on her bed to all hours of the night working on matters relating to the house and my father's work. She always had a bell by her bed in case she needed help, and was prone, as I recall, to nightmares where she would have to be woken up from eerie cries.

My mother felt safe with Richard in the house, but she more than once twisted her hands in a motion of wringing a chicken's neck when talking about him. He complained of her television blasting below him all night and generally felt unappreciated. Jan, who returned to help my mother, each morning greeted her with a mumble, she complained, reminding her of her loss of hearing, which she resented.

When my mother died, Shulamit had a vendetta against Richard coercing him to move out of Hartley. Probably his German descent didn't endear him to my sister, and she repeatedly showed me proof of his disloyalty and incompetence.

My sister also got rid of Jan. He simply had too many interests and when unsupervised he worked on something unrelated. His interests and conviction that Velikovsky was right both diminished. My father nick-named Jan "lightning rod" – grounding my father's anger.

As we saw in the case of Mr. Marx, not all of Aba's supporters were genuine. Marx was a German who created a chronology map based on Aba's book, *Ages in Chaos*. Impressed with his work, Aba agreed to meet him. Mr. Marx then catapulted into our lives causing havoc, and became a particular thorn in my mother's side. Arriving in Princeton, he wanted to take charge of my father's work. Learning to regret his associa-

Aba in Deerpath Pool

On Lavalette Boardwalk in N. J.
Photo by Leroy Ellenberger

On Tel-Aviv Beach

tion with Marx, Aba, on several occasions, slapped himself on both cheeks in a statement of regret for not having listened to me. Many letters were composed to Marx over the next few years by both my parents in an attempt to disentangle themselves from a man who became an obsession. After Aba's death, my mother was determined to clarify that Aba did not, as Marx had concluded, explain the Holocaust as part of his theories.

Warner Sizemore, a pot-bellied, non-tax paying reverend who sells his crafts, built up a big Velikovsky archive. Not thankful when my father bought him a washer and dryer for his new home, when my father became depressed, Sizemore turned his back on my request to be of help.

From a robust man, my father became frail the last years of his life, no longer taking big strides, walking behind my mother on the boardwalk, his body having sustained the strains of agitated depression.

He was a vain man concerned with his appearance. Whenever he came to our house in warm weather, my mother would encourage him to take a dip in our pool. Standing by the pool in his trunks, he'd inhale, proudly expanding his chest, then, looking pleased, stepped into the cold water and quickly squatted, the water reaching his neck, getting admiration from onlookers for his courage. He held onto the rope and kicked his legs. My mother avoided the sun and sat in the shade with a straw hat. Her skin having infrequently been exposed to the sun, remained that of a young woman. She was happy when my father was relaxing.

McMaster University, June 19, 1974
Photo by Wallace Thornhill

At his last lecture at Princeton University, Aba received a standing ovation. His voice was weak – he could barely be heard. He talked to a standing room audience. My mother, next to me in the front row, concerned he could not be heard, requested the microphone be placed closer. She frequently motioned to my father during his lectures not to go off on tangents. He had much information, little time, and was eager to teach. Going off on tangents was not the problem, returning to the original point was. I found it difficult to deal with pockets of whispering skeptics and with those who, when Aba prolonged the question-answer period, left. At this last lecture, his mind was clear, but his voice was weak.

Aba's Death

My mother, a worrier, was always relieved to hear that everyone in the family was alright. She would have a "heavy heart", as she called it, when someone was ill. She would put down her violin and bow intermittently during playing chamber music to check on my father. She was happy when I spent time with him. I saw my parents almost daily and talked to them on the phone several times a day.

The morning before Aba died, standing in the doorway of the second floor bathroom, I watched him shave with a razor. He glanced at himself in the mirror as he was shaving, an image I repeatedly recall. He appeared young and healthy. We then sat at the round kitchen table and he ate a bowl of oatmeal prepared by my mother who had gone downstairs to play chamber music. He suddenly looked old and depressed. My mother was glad I was there to spend time with Aba.

To his Israeli grandchildren he was "Daddy", their own father being "Aba" ("father"). My mother occasionally called him "Daddy." Preferring not to be referred to as grandparents, my father jokingly told of a man who exclaimed that it was O.K. to be a grandfather, but to be married to a grandmother, was another matter! His followers mostly called him "Velikovsky."

While eating the oatmeal that morning, Aba said something about being ashamed to face his neighbors on Hartley because Sagan had given him the reputation of being a charlatan. We went downstairs, he put on his heavy winter coat and scarf and we went for a walk. We crossed Hartley Avenue and walked along the newly constructed university housing. We walked down a few steps and then along a winding road by the retired professors' homes.

It was cold and he worried that I had no gloves on. Passing by an orange berried bush, I gave him a pep talk, and asked, "What is the alternative (to life)?" Encouraged, he said something about going to the seashore for the weekend. We returned to the house. As he stepped up onto the slate platform, perhaps twelve inches from the driveway, he did so with effort. Since then, every time I took that step, I thought of my father's step, and it took effort. We went into the house and he sat on the couch in the room adjoining the music room and listened to the music as he ate a banana.

I returned a couple of hours later and my father, wearing a heavy Norwegian sweater, was lying on his bed. He suggested we go out to lunch to the Nassau Inn, and asked my mother to call and find out what was on the menu, since Richard, the young man who had arrived three days earlier to work for my father, was a vegetarian. My father was lying on his side facing the wall, and it was apparent that, having often sacrificed himself for his family, he wanted to go to lunch just to please me. He seemed tired. We did not go. I went to my office in New York.

I was terribly concerned with a medical report he had received a few days earlier which had frightened him and convinced him he had prostrate cancer.

When I returned in the evening from New York, my husband told me that my father was not feeling well and that a doctor was coming to the house. I rushed over there. My father's room, which he himself had paneled years before, was neat. The layers of pants usually draped over the back of his chair and piles of newspapers and notebooks were all gone.

He complained of an infection in his big toe that would not heal. The doctor arrived and said my father's blood pressure had dropped severely. He was cold. Having a recent lab report in front of him, the doctor still did not conclude that the diabetes was out of control. At first, the doctor considered hospitalization and my father concerned himself with kosher meals in the hospital. My father talked to the doctor about my mother's aneurysm, saying in essence, that she was the sick one. When the doctor heard that my father had discontinued taking medication and that once before he had had a similar reaction to termination of medication, he decided that Aba should stay at home. That was his death sentence. The doctor left. My father, looking angry, asked me to get Ima. He asked her for a bedpan. I left the house never to see him alive again.

My mother later told me she had called the doctor at midnight. Again, my father was not hospitalized. My mother moved a cot into Aba's room and slept next to him. She pleaded for a moment's sleep for which she later felt guilty. He got up and took a shower, and then took his pillow wandering through the house looking for the couch at the beach house. He was disoriented. A doctor later explained the disorientation was due to the diabetes flare-up.

Early in the morning my father complained of a sharp pain in his back. An hour or two later my mother called me and I jumped up to rush over there. It was about 8 a.m. and she said, "Aba is not feeling well." She put him on the phone – I did not know why. I said, "Aba! Aba! Aba!" He did not reply. She took the phone and said he had died.

I called the Rescue Squad. Unsuccessful in reaching them, I drove to the Rescue Squad on Harrison Street and rang their bell.

The contrast between life and death was sharply crystallized. The Rescue Squad arrived and left as quickly as they rushed in. There was no one to rescue. The doctor arrived – I glared into his grey eyes and asked him if my father was dead. The doctor asked why Aba was on the floor. The Rescue Squad had placed him there to resuscitate him. An eighty-four year old man of great dignity dead on the floor of his room.

My mother was calm. She called the New York newspapers as though she had long ago prepared herself for that moment. The obituaries, it seemed, had also been prepared long in advance.

The treating physician, who did not arrive that weekend, had given him an appointment two weeks later, even though my mother called repeatedly. The doctor had frightened my father about prostate cancer, and meeting me unexpectedly at the local bank, he blurted out that my father had prostate cancer and that his testicles would have to be removed. I had made several calls that week to my father's urologist in New York who said that the Princeton doctor was frightening my father unnecessarily and was driving my father to despair with an unsubstantiated diagnosis. I made many calls to put the physicians in touch with each other. They talked – but Aba died two days before his scheduled appointment.

I was devastated. I called my husband, my son and my daughters who had heard the Rescue Squad sirens go by. When three stars appeared in the darkened sky, signaling the end of the Sabbath, I called my sister who was in Denver at the time. She concerned herself with a proper Jewish burial. Aba was a saint, she stressed, for he had died on the Sabbath.

I woke up many nights in deep despair. One night I dreamt my father's soul was spinning faster than the speed of light into the universe.

Aba had said to me that when he died I should not let his religious Israeli granddaughter arrive and make a fuss. He said he wanted to be cremated, but did not want to upset my sister. It was my mother's decision to have a graveside private ceremony. I did not want those who disrespected Aba during his life to make an appearance at the end. His books had made him a public figure and in death he was, once again, ours.

The sun was shining on the small secluded cemetery in Tom's River. Before internment my mother made certain my father's open eyes were bandaged – a religious dictate. My mother, my husband, my three children, and I held hands and cried as the coffin was lowered. Returning to Princeton we stopped at a restaurant for which my mother later felt guilty. She repeatedly said she should have saved my father.

Sometime later, at a Velikovsky symposium at the Princeton Holiday Inn, my mother lit a candle and recited my father's words:

> I was compelled by logic and by evidence to penetrate into so many premises of the house of science. I freely admit to having repeatedly caused fires, though the candle in my hand was carried only for illumination.

I was reminded of how much Aba had suffered and that I was no longer able to talk with him.

Ima at the Helm

The first pages of Velikovsky's notebook

Having been first to invent the electroencephalogram, having theories about making the blind see with electricity, ideas about curing cancer, Aba left much unfinished work. He noted that Eskimos had a low frequency of cancer and wondered about their climate and diet. He had ideas about the origin of religion and wanted to write a book about Jesus. He knew where oil could be found around the world, and his hand written notebook revealed many unfinished projects.

In spite of her own acclaim as both a musician and an artist, Aba was the core to my mother's life. After he died she told me nostalgically how she loved to lie next to him with her head on his shoulder. I had never seen a display of physical affection between my parents, save a kiss my father rarely planted on her forehead. There were other expressions of his affection, such as giving my mother flowers on my and my sister's birthdays. My father, having undergone stomach surgery while on a visit to Israel, sent flowers to my mother from the hospital. My father treated her like a queen.

For Aba's autobiography *Days and Years* (which is still to be published) Ima wrote the following page:

> When in 1940, in the early evening after a day in the Public Library at Forty-second Street in New York Emanuel and I walked to Central Park and sat down on a bench, I said to him life was not always easy with him, but it was certainly an adventure. This sums up in some way my life with Emanuel. In this autobiography of his this feeling of adventure comes through. Almost every day, especially since we came to the United States, unusual things happened.
>
> He was always on a hunt for more knowledge, following roads into different directions to look for and find more clues to his intuitive thoughts and expectations. He was like a hunter on a trail – though he would never have hunted an animal or gone fishing, he respected life too much for such sports – but as a man of a vision, who looked into many directions, he was driven by a never ending urge to know more.
>
> He was humble and proud at the same time, and before all he was a great fighter who never took no for an answer; who went to the authorities of the past as well as to the great living to ask for explanations. He was never discouraged. When one of his expectations was not fulfilled, he went into the next direction to find the answers he looked for. He abandoned one way and went to the next without losing confidence in his search for what he hoped to be the solution and the truth. All in all, Emanuel was a man of a very unusual character. As I said, proud and humble, courageous, never to give up even when the odds against him and the personal attacks on him would have overwhelmed many a man. Of course, the strain showed through his life. There were times when he felt depressed as in reaction to the difficulties found in his way. And no wonder, I lived with him through very trying times and I understood when the load became too heavy.

cont'd

He was a very gentle human being, he cared for his parents, family, friends, and for the plain and simple. He lent a hand to people, he found time to comfort and advise and help, never being too busy to listen to the unfortunate, adults and children alike. He was a great optimist who believed in the goodness of man and in the purpose of life, and he also had a fine sense of humor. He was a great raconteur and would tell anecdotes and stories. I would ask him to cheer me up when I felt let down, and he would say, let us count our blessings. And we would count. And that helped every time to make us feel positive and happy again. He gave this advice also to his friends, who still remember him for that. It was always "the cup is half full and not half empty." He called me Shevik, and when he was in a sentimental mood I was Shevinka; but when he called me Elisheva, I knew that something was wrong, that he did not approve of me at that moment. I, however, never had a nickname for him—he always was Emanuel for me, also Aba. Some of his family called him Monia, but to me it sounded too much like "money," and this association did not fit him at all. He could have gotten rich many times, but it was not written in his stars, and certainly not in mine. And when he was sometimes sorry that he did not take advantage of an opportunity to buy land and enrich himself and his family, I would say, never mind, you bought the sky. And he would smile.

He wrote about his first 6½ years in several but similar versions and in several languages: Russian, German, Hebrew and English. He made lists how to divide his life into different epochs, but wrote only sporadically, and some parts are missing. There were also letters, correspondence with his father, with friends and scholars, which are biographical. There were no letters between us, because we were almost never separated. All the trips to Europe and Palestine we made together, and I went with him across the United States to campuses to be with him when he lectured.

cont'd

A few sentences of friends after his death:

S. Vaughan wrote me a poetic line: "So Velikovsky has left this fragile ship, but he is sailing the seas of space."

Walter Kaufmann quoted Fulton Oursler: "Do not mourn that he has died, be glad that he has lived."

And a recent note to me from a reader I don't know: "May I ... tell you how admiring I am of your matchless memorial to this great man – the publication of his unique works." And that is what I have been doing these past two years, and that is what sustains me.

There is a saying which dates back to older times – Herodotus records it of Solon speaking of Croesus, the richest man, whether he was lucky. Only after his death do you know whether a person was lucky.
In this sense Emanuel was a lucky man. On a Sabbath morning, lying quietly on his bed after a rather restless night, speaking softly to me – I was sitting next to him, on the edge of his bed, and touching his shoulder asked him to repeat his last words I did not hear clearly; he turned his head a bit to the side, he did not answer – he had died – without a gasp, without a murmur. He was a lucky man in many ways because of his strong character, his honesty with himself, his total devotion and integrity, a man of vision, of commitment with belief in his work.

Elisheva Velikovsky

Dear Lynn:

I repeat here
what I said so often.
Should I become incapacitated
or die, you together with
Elisheva should have the
authority to decide what to do
with my archive, and also
with single manuscripts, and have
a free access to all my
literary material and the
right with Elisheva (when she
is no more alive, with Ruth
Sharon)
to act as literary executors.

Yours,
Im. Velikovsky
Princeton, Aug. 1, 76

Jan Sammer, witness

Aba's death opened the way to the editing and publishing of his final works. He was out of the way and no longer controlled the publication of his works. Whether by design or by accident, he had been a major stumbling block discouraging others from taking charge. When he died, Ima was finally able to accomplish what she had wanted to for many years – get his completed manuscripts to the publisher.

As life's irony plays its own game, the posthumous volume, *Mankind in Amnesia*, edited by my mother, was soon to be remaindered. The volume dedicated to the five grandchildren expressed the hope that nuclear holocaust could be prevented. The book was the psychological approach to my father's theories. It called to consciousness how man plays with the atom to re-live the trauma experienced by our planet in historic times. The book sold poorly and went unreviewed.

The second posthumous book, *Stargazers and Gravediggers*, the Watergate of science, published in 1983 when my mother was very ill, was also remaindered. After months of energetic work editing the book, my mother, having had a stroke, took one look at the first copy, put it aside never to look at it again. The book described the history of the reception of *Worlds in Collision* leaving out both torment and glory.

When the great procrastinator, also the founder and instigator of all the excitement, was to die and she could take matters into her own hands and finish what he couldn't, my mother did not reap rewards. Her reward was in the work itself. Every Friday night she would come to our house, a five minute drive from her home, and stay until Sunday. Reclined on her bed, leaning against many pillows, her knees up to support a manuscript, she often exclaimed, "It's great!" referring to my father's work, which she loved. Because she did not write on the Sabbath, she secured paper clips to pages of manuscripts and was later able to recall what the clip positioned along a typed line represented. However, as it turned out, she left manuscripts with paper clips secured to many pages when a stroke to the left side of her brain ended the communication and energy of this amazing woman.

On her 87th birthday, her last, she was given many gifts by my children and me. Embarrassed, she giggled as Professor Lynn Rose at the kitchen table, where she unwrapped the presents, admitted that he knew her age.

Ima died one month short of her 88th birthday. The last ten months of her life were bitter. The matriarch had collapsed into a state of helplessness.

On Labor Day, 1982, I received a call from the Medical Center of Princeton telling me that "… Mrs. Velikovsky has suffered a stroke." I remember thinking it was not possible. I wished I could turn the time back before that call.

Ima had been at the Pelican House for the weekend – the first Saturday she didn't spend with us since my father died.

She spent a pleasant evening with Rafael and his friends who were visiting. In the morning, after Rafael had gone to the beach, she complained of difficulty moving. Jan drove her to Princeton. On the trip her speech became slurred. Lying on a cot at the emergency room, she smiled at me, and only one side of her face smiled.

I had not known what a stroke could be like. A particularly debilitating condition in the arsenal of illnesses, it effectively ended the vibrant person I had always known as my mother. From that day, with the exception of six weeks at the Columbia Presbyterian Hospital and the last month at the Medical Center of Princeton, she spent in my home. Traveling by ambulance from Princeton to New York on the second day of her illness, seeking the best medical care, sitting next to her, I explained to her each turn and where we were.

The Columbia Presbyterian Neurological Unit was dark and depressing. Across the hall music played in a room where a comatose old woman was fed by tubes. Taking care of my mother, scrutinizing round the clock private nurses, calling doctors, talking with her in between seeing my own patients, visiting her and having sessions with other patients over the hospital lobby telephones – until I could not move – my life had become a preoccupation with saving my mother.

I took charge of my mother's care, supervising the occupational therapists, the running of the house, the help and nurses and administering medication together with running Hartley, where Richard Heinberg and Jan Sammer, who were working for my mother when she had a stroke, lived and cataloged work left by my father as well as copying cassettes of my father's lectures.

Shulamit, my sister, arrived from Israel while my mother was hospitalized in New York and was very loving, very helpful and very concerned. Upon her discharge, my mother asked to stay in my home and have Dr. Seed in charge.

When she began to realize that the stroke was irreversible, for the second time in her life she went into a deep depression. She pointed to the foot of her bed and cried, turned over the food-ladened spoon to examine it from all sides. She was hallucinating. Dr. Seed made house calls every night; sometimes twice and even three times a day when my mother pulled out the feeding tube from her nose.

Reading medical books, my sister took issue with the medication prescribed for my mother's depression and told my mother not to take it. Tension built and my sister returned to Israel.

Immediately after she suffered the stroke and for weeks afterwards, my mother was able to say a few words, repeating "Aba did it the right way!" – my father had died quickly and relatively painlessly. As things turned out she was diagnosed as also having cancer. An autopsy revealed a wide spread cancer in the intestinal area which destroyed her immune system (and allowed a breast cancer two years earlier, treated with a mastectomy) and finally cancer of the bladder and a fungus infection in the blood. Not treating the blood infection would have meant sure death within ten days. Treating it meant either cure or the malfunctioning of her kidneys. I wrenched at my conscience and cried bitterly – eventually opting for the treatment – at least there was a chance. I asked her in the intensive care unit (in a doctor's presence) if she wanted to live and she blinked "yes" to life.

My mother was a vain woman interested in looking her best and in receiving compliments. She particularly liked Rafael's compliments as he planted a kiss on her cheek upon greeting her (to the last days of her life), and while well, telling her how nice she looked. Confronted by her mirror image after she had the stroke was painful, and she covered her face with both hands.

The realities of the stroke are even uglier than one could imagine. The first day the neurologist told us that my mother would not be able to communicate again. How was that possible? This vibrant woman who lived alongside this difficult, loving giant through a lifetime of turmoil, would not be able to overcome this tragedy? I expended great effort – providing and monitoring speech therapists, physical therapists, etc. – but all for naught – my mother wanted none of it. She kicked away the physical therapists and went to sleep for the speech therapists.

She sat at the piano, this musical woman, and played the same three notes over and over, walked a few steps if led and supported, tried to cry but couldn't and occasionally called my name. Depleted by life's toll, she occasionally bellowed in despair. My hopes and determination for her cure were undampened. Eventually, she began to enjoy television and would muster up some laughter. Then she was diagnosed as suffering from cancer of the bladder. She was valiant to the end. No part of her body escaped the indignities of illness and old age. Her dentures did not fit and her lower front teeth cutting her gums were extracted. A mastectomy two years earlier to arrest breast cancer, a tube in her nose for feeding (having stopped swallowing), a catheter to remove urine, and her wig no longer fitting as her hair thinned even more.

Carmel ("Carmelli" to my mother), a special education teacher, patiently and attentively taught my mother to read after she had had the stroke. Naomi ("Numile" to my mother) often visited from East Longmeadow, Massachusetts, and they shared deep closeness.

Two days before my mother died at the Princeton Medical Center, when she was lucid, I called my sister from my mother's bedside, as I had done many times before, so they would communicate. My sister gave her information about her great grandchildren and my mother closed her eyes. Whatever she heard I don't know.

On Friday morning, June 24th, 1983, around 5 a.m., Dr. Seed called to say that my mother had died. Did I want to see her? I did not want to see her dead. I asked Dr. Seed to please make certain that she was really dead, for her fear, of which she often talked and wrote, was being buried alive. When she was born, she told me that she had to be severely shaken to start her breathing, whereby breaking one arm and injuring her neck. She told me that that was the arm that later held the violin, which was also supported by her chin (involving her neck). Also in her twenties she had nearly drowned. She was fearful of being mistaken for dead.

Informing my sister of Ima's death, I received a call a short time later from my brother-in-law in Israel requesting that the doctor make certain my mother was dead. My mother communicated that fear to him also.

My husband, three children, and I stood at the cemetery on that sunny Friday after the autopsy had been performed. Holding hands, we cried as the coffin was lowered.

Life by its very definition, is a disillusionment, for it ends in death. Getting older means that what we hoped for, and which is worse, what we expected for (and not of) our children mostly fails. Instead of getting more pleasure as we accumulate friends and possessions, we spend more money for which we get little, we do not look as well, and we do not feel as well. We lose our parents and realize that one day we too will die, and hope to avert the death (of my mother's) which we viewed as most horrid – the slow, silencing stroke coupled with cancer and depression. On the other hand, the way my father died, so very suddenly, left more guilt, for there was no time to help him. However, he spared us the months of anguish and torment of racing to save a failing body with which we cannot part.

Aba's Legacy

Department of Rare Books and Special Collections
Princeton University Libraries
Princeton, New Jersey 08544

DEED OF GIFT

Ruth Velikovsky Sharon, 50 Deer Path, Princeton, New Jersey 08540, hereinafter Donor, donates to the Princeton University Library the papers of her father Immanuel Velikovsky (1895-1979). The paper are hereinafter referred to as "Collection."

The Collection includes the following materials:

- Immanuel Velikovsky papers deposited (deposit no. 9532) in the Princeton University Library (70 document boxes and cartons; 1 container of film)
- Immanuel Velikovsky papers in the Donor's home, including all Albert Einstein handwritten or typed/signed letters, Einstein handwritten notes and comments on letters of Velikovsky, 1 Sigmund Freud post card, 1 Carl Jung letter, and 1 Eugen Bleuer letter
- Any other original Immanuel Velikovsky papers, such as correspondence, manuscripts, and other items in the Velikovsky papers in the Donor's home, especially such original items as are currently represented only by photocopies in the Velikovsky papers deposited at Princeton.

Note: The Donor will retain three Einstein letters dated 27 August 1952, 22 May 1954, and 12 March 1955; and a Sigmund Freud letter. An Einstein letter dated 2 January 1951 and a poem dated 15 June 1956 cannot be located currently; if found, they will be given to the Princeton University Library.

---

1. Property Rights: The Collection is a gift and upon transfer to the Library becomes the property of The Trustees of Princeton University. The Einstein, Freud, and other autograph materials enumerated above will be transferred as of 31 July 2004. Additional papers of Immanuel Velikovsky at the Donor's home will be transferred as of 31 August 2004. In accordance with IRS regulations, the Donor is responsible for the cost of an independent professional appraiser's written appraisal for tax purposes.

2. Copyright: Copyright and literary rights will be retained by the creator of the materials in the Collection, or by their heirs and assigns, in accordance with prevailing U.S. copyright law. The Donor and her heirs or assigns will retain Immanuel Velikovsky's copyright and literary rights in manuscripts, letters, and other textual materials that he authored.

3. Access: Once the Collection has been arranged and described by the Library, it will be open to all researchers, including all members of Immanuel Velikovsky's family, in accordance with Princeton University policy on fair and equal access to information. However, in order to protect the Donor's copyright, no more than 250 exposures in total will be permitted either to any individual researcher who requests copies, or for any particular purpose such as publication of Velikovsky's works and correspondence, either in print or on a website.

4. Disposition: If there are duplicates and other materials inappropriate for permanent retention in the Collection, they will be offered back to the Donor; if not wanted by the Donor, such materials will be transferred or disposed of in accordance with the Library's usual policies and procedures.

Donor/Legal Agent: _Ruth V_____ Sh_____
                       Name   Title

Date _June 16, 04_

Recipient _Don Skemer_ Curator of Manuscripts
            On behalf of Princeton University   Title

Date _June 16, 2004_

Once my mother died I had no personal reason to prove Aba right anymore. She was eager to get her husband recognition. I cared little after my mother's death what anyone thought of my father's theories – until a nagging feeling of responsibility to my three children, who are included in the dedication to *Mankind in Amnesia,* prompted me to give this personal account of my father's life. The need exists to trigger interest in his work – that, as I understand it, will touch the fabric of American society.

The censoring of *Worlds in Collision* is described in *Stargazers and Gravediggers*, incriminating some of the "great" men of science. For years the incomplete manuscript for *Stargazers and Gravediggers* was on my father's desk. He did not publish it because my mother, who was a peace loving woman, feared litigation. Ironically, after my father died, my mother arranged for its publication. By the time the final galleys arrived, she had had a stroke and had no interest in the book.

In 2004 I donated my father's collected papers to Princeton University Library, in order to make them available for research use. The collection comprised 156 boxes of published and unpublished manuscripts and drafts, lectures, correspondence, and legal documents.

For further reference, please visit:
http://arks.princeton.edu/ark:/88435/v692t621n

PRINCETON UNIVERSITY

H. KIRK UNRUH, JR. '70
Recording Secretary

December 16, 2004

Ms. Ruth Velikovsky Sharon
50 Deer Path
Princeton, NJ 08540-4036

Dear Ms. Sharon,

Thank you most warmly for your Gift-in-Kind to the Library of a collection of books and miscellaneous papers of Immanuel Velikovsky.

The expansion of our library and its programs over two and a half centuries is one of the great Princeton stories and reminds us how fluid a thing the library must be, and how it constantly renews and adapts itself to changing times. Your gift enables us to keep it strong and growing.

With best regards and best wishes for a happy holiday season and a wonderful new year.

Cordially,

H. Kirk Unruh, Jr. '70

HKU: kjc
20190406

News at Princeton (printed from www.princeton.edu)

Library acquires papers of scientist and author Velikovsky
by Ruth Stevens · Posted July 29, 2005; 03:06 p.m.

The papers of Russian-born American scientist and author Dr. Immanuel Velikovsky have a new home in the Princeton University Library. His daughter, Ruth Sharon of Princeton, has donated the papers for use by researchers.

Velikovsky, who lived from 1895 to 1979, is best known as the author of a number of controversial books, primarily arguing that ancient myths, legends and accounts of catastrophic events related in the Bible and other texts have a basis in fact.

Velikovsky earned his M.D. degree from the University of Moscow in 1921 and lived in the 1920s and '30s with his family in Palestine, where he pursued a specialization in psychoanalysis and psychotherapy. He moved to the United States in 1939 and began his research on the history of Egypt, Greece and the Jewish past. He lived first in New York City and later in Princeton.

Velikovsky is the author of "Worlds in Collision" (1950), "Ages in Chaos" (1952), "Earth in Upheaval" (1955), "Oedipus and Akhnaton" (1960), "Ramses II and His Time" (1978), "Peoples of the Sea" (1977) and other controversial studies of ancient Egypt, Greece and Biblical history.

"Decades of bitter debate resulted from his controversial theories about cosmological catastrophes, particularly in regard to the planet Venus, and his interpretation of historical, geological and paleontological evidence," said Don Skemer, curator of manuscripts in the library's Department of Rare Books and Special Collections.

Harlow Shapley, Cecilia Payne-Gaposchkin, Carl Sagan and other members of the American scientific establishment were among Velikovsky's critics. The controversies over Velikovsky and publication of his work have been the subject of several books, including Alfred De Grazia's "The Velikovsky Affair: The Warfare of Science and Scientism" (1966) and Donald Goldsmith's "Scientists Confront Velikovsky" (1977).

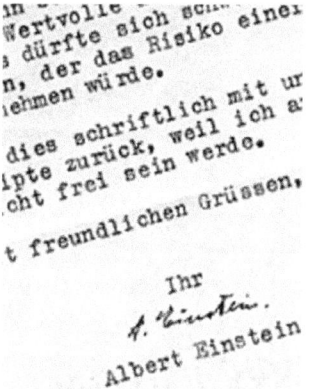

The papers donated to the Princeton University Library include a July 8, 1946, letter (below) that Albert Einstein wrote to Immanuel Velikovsky (above) after reading an early version of what became the book, "Worlds in Collision" (1950)

The Immanuel Velikovsky Papers comprise 156 boxes of published and unpublished manuscripts and drafts, subject files, lectures, personal and professional correspondence, and legal documents dating from the 1920s to the 1990s. The collection includes his extensive correspondence with Horace Kallen, Ted Thackrey and other friends, as well as selected letters from Albert Einstein, Helen Dukas, Sigmund Freud, Carl Jung, Eugen Bleuler, Chaim Weitzmann, Harold Urey and other scientists and academicians.

His papers also contain extensive correspondence with publishers and readers. Among Velikovsky's manuscripts are selected typescript essays annotated by Einstein, which later became chapters in Velikovsky's posthumously published book, "Stargazers and Gravediggers: Memoirs to Worlds in Collision" (1983).

The Immanuel Velikovsky Papers are available for research use in the library's manuscripts division in the Department of Rare Books and Special Collections. An online description of the papers is available. For more information about the papers, contact Skemer.

June 17-1965

To the editor of the New York Times:

Dear Sir:

What would we give if we could We are today tell Mozart how greatly indebted us, and to to him for all the pleasure he has given generations, & forever after, with his great music.

What would we give if we could thank Galileo today for all his great contribution to both science and independent thought. Also to apologize for the treatment he received from his contemporaries — for we know today what a great scientist he was. A children's encyclopedia simplifies the reasons for Galileo's troubles: "Galileo made himself unpopular because he dared to doubt the learned people of earlier times. He was even imprisoned and threatened with death for some of his beliefs. But he accomplished a great deal. He started scientists to finding out for themselves new things about the world around us."

About Copernicus the encyclopedia says: "His ideas seem so simple to us now that it is hard to understand why people fought so hard against them. He did not live to see the great storm his book stirred up."

Letter 20
A letter to the publisher of the New York Times, I had written in 1965

although there are many more.

My cry is as a daughter. What can I do for my most beloved father. True, he has many supporters and followers — files of letters of great praise. But are we going to let one more great man go to his grave without clear-cut recognition, with the continuous slander of the boycotting scientists who cannot be rational.

On the campuses the young minds are grasping his greatness. And from Brown University comes the following quotation (in their newspaper).

"The unscientific fury and vituperative reactions that have descended upon him are almost irrational. "If Velikovsky makes sense, why not test his theories? If Velikovsky is 'too ludicrous to merit serious rebuttal,' then why did the whole of science rise up in righteous indignation against him? These questions are pertinent."

I as a daughter have felt pride mixed with anguish. At my father's insistence I have set idely by. He feels that the tide has changed and in my lifetime I will see his name among the honorable great men. I have no patience. In my lifetime it is too late. My father just turned 70. And I pray that in this free country where free thought is encouraged, your paper will come forth and print my plea.

Sincerely yours.
Ruth Sharon   50 Deer Path, Princeton, N.

Letter 20

June 17, 1965

To the editor of the New York Times:

Dear Sir:

What would we give if we could today tell Mozart how greatly indebted we are to him for all the pleasure he has given us and to generations, and forever after, with his great music.

What would we give if we could thank Galileo today for all his great contribution to both science and independent thought. Also to apologize for the treatment he received from his contemporaries – for we know today what a great scientist he was. A children's encyclopedia simplifies the reason for Galileo's troubles: "Galileo made himself unpopular because he dared to doubt the learned people of earlier times. He was even imprisoned and threatened with death for some of his beliefs. But he accomplished a great deal. He started scientists to finding out for themselves new things about the world around us."

About Copernicus the encyclopedia says: "His ideas seem so simple to us now that it is hard to understand why people fought so hard against them. He did not live to see the great storm his book stirred up."

...

My cry is as a daughter: What can I do for my most beloved father. True he has many supporters and followers – files of letters of great praise. But are we going to let one more great man go to his grave without clear-cut recognition, with the continuous slander of the boycotting scientists who cannot be rational.

On the campuses the young minds are grasping his greatness. And from Brown University comes the following quotation (in their newspapers):

"The unscientific fury and vituperative reactions that have descended upon him are almost irrational. If Velikovsky makes sense, why not test his theories? If Velikovsky is "too ludicrous to merit serious rebuttal," then why did the whole of science rise up in righteous indignation against him? These questions are pertinent."

I as a daughter have felt pride mixed with anguish. At my father's insistence I have sat idely by. He feels that the tide has changed, and in my lifetime I will see his name among the honorable great men. I have no patience. In my lifetime it is too late. My father just turned 70 – and I pray that in this free country where free thought is encouraged, your paper will come forth and print my plea.

Sincerely yours,

Ruth Sharon, 50 Deer Path, Princeton, NJ

Auburn University, Auburn, Alabama, April 1975

Immanuel Velikovsky's Interdisciplinary Synthesis

In *Stargazers and Gravediggers: Memoirs to Worlds in Collision*, Velikovsky discusses his theories very briefly, then writes:

> Should I go on? Should I make the mistake of digesting my book and enabling still more people to discuss its merits and demerits, knowing it only from a condensation? I cannot compress *Worlds in Collision* any more than it is in its present form as a book – there I have not left a sentence that I deemed superfluous. (*Stargazers and Gravediggers*, File I, »Worlds in Collision«)

This paragraph reveals my father's dislike for condensations or summaries of his work. Hence, it should be pointed out that all of the evidence on which his theories are based is not presented here. It is impossible to present the extensive detailed arguments and the data to support those arguments in an article or brief summary.

The summary that follows is meant to acquaint the reader with Velikovsky's theories and to present examples that illustrate his method of integrating data from different disciplines to arrive at an interdisciplinary synthesis. For a thorough understanding of Velikovsky's theories, the reader should carefully read his books, *Worlds in Collision* (1950), *Earth in Upheaval* (1955), *Ages in Chaos* (1952), *Peoples of the Sea* (1977) and *Ramses II and His Time* (1978). Posthumously published: *Mankind in Amnesia* (1982) and *Stargazers and Gravediggers* (1983).

According to Immanuel Velikovsky, global catastrophes took place on earth during recent history and within human memory, and the cause of these catastrophes was extraterrestrial. There were shifts in the position of the planets, with Venus being the primary agent.

During the eighth and seventh centuries B.C. the planet Venus appeared for the first time, formed from a comet, which had earlier been ejected from the planet Jupiter. This comet then approached Mars, which, in turn, nearly collided with the Earth around 687 B.C. Twice, the Earth passed through the comet's tail. After this, the comet (the protoplanet, Venus) lost its tail and settled into the solar system as Venus, the planet.

The effect of these interplanetary interactions on Earth was catastrophic on a global scale. Oceans were displaced, mountains collapsed while others were created, continents were flooded with enormous tides, there was intense heating and incessant electrical discharges, and volcanoes around the world erupted spewing out lava. Hurricanes swept our planet, hot stones, or meteorites showered all parts of the globe, a sticky liquid, naphtha, rained down, the result of some hydrogen and carbon gases in the tail of the comet that caused fires to spread over the Earth's surface. There was precipitation of a carbohydrate substance, likely to be the manna upon which the Israelites fed. A shroud of darkness enveloped the Earth from the fall-out from interplanetary dust and lava dust in the sky. Accompanying this widespread destruction were infestation of vermin and plagues of boils and diseases.

These global catastrophes resulted in changes in the Earth's orbit and the tilting of its axis, the magnetic poles reversed. Changes occurred in the Moon's orbit, and in length of the day, seasons and year. Between the fifteenth and eighth centuries B.C. the length of the year was 360 days, but starting in 687 B.C. the year increased to 365 ¼ days.

This reconstruction was carefully completed after ten years of thorough, detailed cross-cultural research and analysis of the texts of ancient writings from all civilizations. This includes observations of the solar system by ancient astronomers, memories recorded in myths and legends, and obsolete calenders and sundials. He also looked to evidence in geology, paleontology, astronomy, and physics.

Velikovsky saw striking similarities among ancient writings that point to a universal experience of catastrophic events. When he first read the Egyptian *Ipuwer Papyrus*, he realized it looked like a copy of the *Book of Exodus*. And, when he researched further, he saw the same events recorded in the ancient writings of Mexico, South America, Scandinavia, Polynesia, Africa, the Laplanders, the Eskimos, Asia, Indonesia, and on and on.

With his revolutionary interdisciplinary perspective, he also looked to the physical and natural sciences to reconstruct these events. Now, more than half a century later, scientists are using Velikovsky's approach and presenting precisely the same conclusions, but without credit being given to Velikovsky. The following examples, only a handful of hundreds in his books, illustrate his integrative theoretical perspective. The reader, however, should refer to his books, to see the vast quantity of evidence he offers.

For example, one of the effects of this catastrophe was the reddening of the Earth's surface, both land and sea, by a "fine dust of rusty pigment." This phenomenon was observed and recorded by the Israelites, the Babylonians, the Egyptians, the Finns, the Central Americans and other civilizations.

The Mayans wrote that "The sun's motion was interrupted, the water in the rivers turned to blood." Ipuwer recorded that "The river is blood," and this corresponds to the *Book of Exodus* that says "All the waters that were in the river were turned to blood." The Red Sea is likely to have received its name from this phenomenon, asserts Velikovsky. "In one Egyptian myth the bloody hue of the world is ascribed to the blood of Osiris, the mortally wounded planet god; in another myth it is the blood of Seth or Apopi; in the Babylonian myth the world was colored by the blood of the slain Tiamat, the heavenly monster."

Drawing also in specific literature, Velikovsky explains that "The fall of meteorite dust is a phenomenon generally known to take place mainly after the passage of the meteorites; this dust is found on the snow of mountains and in polar regions." (see *Worlds in Collision*, Part I, Ch. 2, »The Red World«)

Another example is the precipitation of naphtha and the resulting fires. These are described by the Maya as "... a great inundation ... people were drowned in a sticky substance raining from the sky" and by ancient Siberian Voguls as a " ... sea of fire upon the earth ... The cause of the fire they called the fire water." The aboriginal people of the East Indies say the "'water of fire' rained from the sky; with very few exceptions, all men died." The *Book of Exodus* tells of a "fire that ran along the ground", the Egyptian *Papyrus Ipuwer* describes this consuming fire: "Gates, columns, and walls are consumed by fire" and the *Midrashim* say that "The Egyptians refused to let the Israelites go, and He poured out naphtha over them, burning blains (blister)." (See *Worlds in Collision*, Part I, Ch. 2, »Naphtha«)

With this historical and cross-cultural data, Velikovsky integrated archaeological and geological evidence. He discovered that the above civilizations today actually have petroleum deposits (Mexico, Siberia, Iraq and Egypt). And, a tomb of a pharaoh of the Middle Kingdom shows evidence of the effects of a raging fire from a light combustible material that left no charred remains.

Evidence of a prolonged period of darkness also exists cross-culturally, many having spoken of "a cosmic catastrophe during which the sun did not shine …" The Finns wrote that "hailstones of iron fell from the sky, and the sun and moon disappeared …" the Indians of the New World recorded that the sun did not appear for five days; a cosmic collision of stars preceded the cataclysm …," the Babylonians wrote about a dark cloud, "Desolation … stretched to heaven; all that was bright was turned to darkness. …" Hebrew and Egyptian texts refer to the same period of darkness. This darkness Velikovsky attributes to lava dust from the simultaneous and prolonged eruption of thousands of volcanoes and the comet's dust (See *Worlds in Collision,* Part I, Ch. 2, »The Darkness«). In recent years, Carl Sagan, Velikovsky's major critic, echoes Velikovsky's findings, without credit to Velikovsky, when he writes about nuclear winters.

Velikovsky conducted extensive research with ancient astronomical records, which suggested that the Babylonians and Hindus did not see the planet Venus, despite Venus being the most conspicuous one. The Babylonians spoke of a four-planet solar system – Mars, Mercury, Jupiter and Saturn. Venus was added later. Apollonius Rhodius refers to a time "when not all orbs were yet in the heavens." And, "In an ancient Hindu table of planets, attributed to the year 3102 B.C., Venus alone among the visible planets is absent. The Brahmans of the early period did not know the five planet system, and only in a later ("middle") period did the Brahmans speak of five planets."

The ancients also described Venus as a comet. Spaniards wrote that the ancient Mexicans believed Venus smoked "The star that smoked, la estrella que humeava, was Sitlae choloha, which the Spaniards called Venus." In the *Vedas* it is written about the star Venus Zebbaj or "the one with hair." An ancient astronomer described comets as appearing hairy, "hair or 'coma' is a characteristic of comets, and in fact 'comet' is derived from the Greek word for 'hair'. The Peruvian name 'Chaska' ('wavy-haired') is still the name for Venus." (See Worlds in Collision, Part I, Ch. 8, »The Comet Venus«)

Velikovsky's work had tremendous implications for the fields of astronomy, geology, anthropology/archeology, and paleontology, and evolution. For astronomy and physics, Velikovsky's work shattered the notion of a peaceful static universe. It raised the issue that gravity and inertia are not the only forces at work, that electromagnetic forces must be reckoned with.

For geology, paleontology, and evolution, his evidence cast serious doubt on the uniformitarian doctrine that the Earth's surface has undergone slow, gradual change by erosion and deposition. He helped shed light on the sudden annihilation of some species and proliferation of others, and pointed to extraterrestrial radiation, chemical pollution and electromagnetic forces as some triggers of mutations in the evolution of species. And, his theories are entirely consistent with the theory of punctuated equilibrium, i.e. that evolution involves periods of regularity punctuated by leaps or spurts. *Earth in Upheaval* deals with these issues in great depth.

For anthropology and mythology, his theories suggest that ancient myths, which have been interpreted allegorically, may be based on the witnessing of real traumatic events. And, for psychology, the witnessing of these catastrophes may have been repressed, with a collective amnesia causing humans to re-enact over and over their traumatic experiences through wars, murder, and other acts of violence. Velikovsky devoted an entire book to this topic, *Mankind in Amnesia*.

For ancient history, Velikovsky devoted much of his time to revising the ancient chronology, and published his reconstruction in *Ages in Chaos*. He wrote in *Stargazers and Gravediggers*,

> Since both Egyptian and the biblical time scales are used in the chronologies of other peoples of antiquity, a maze of misconceptions swamped the history of the entire ancient East and had to be disentangled. I worked for more than ten years, strenuously and with enthusiasm, to bring this labor to its completion. (*Stargazers and Gravediggers*, File I, »Ages in Chaos«)

The evidence, of course, speaks for itself. Based on his historical cosmology, Velikovsky made numerous advance claims that were later confirmed with the advent of the space age and continue to be validated until this day. They are too numerous to provide a complete list but some are: hot surface temperature of Venus, the retrograde rotation of Venus, radio noises emitted from Jupiter, the existence of a magnetosphere around the Earth that reaches the moon, a steep thermal gradient a few feet below the lunar surface with heat flowing to the surface, remnant magnetism in lunar rocks and lava (though the Moon itself has hardly any magnetic field), the presence of hydrocarbons, recentness of the last heating of the lunar surface, and the frequent occurrence of moonquakes.

The reader is encouraged to critically review the evidence by reading Immanuel Velikovsky's books. It will be clear that the many recent articles and books on interplanetary collisions, catastrophic events on Earth, and mass extinctions advance the same theories that Velikovsky urged the scientific community to examine for nearly half a century. And, the reader will also become aware of one more thing: Immanuel Velikovsky's name is curiously absent from their bibliographies.

But: Now in 2010 the general respect for my father has multiplied!

Aba at Princeton University
(Photo by Stuart Crump, Jr., Princeton, N. J.)

The Books

(by C. J. Ransom)

This chapter has the purpose of giving a concise overview of my father's books. Although he himself cautioned against summaries of his works (see the preceding chapter), it seems important and necessary today, after decades of having been silenced, to acquaint the younger readers with the topics of the books. After all, the current generation of both laymen and scientists mostly hasn't even heard of Velikovsky and his works, since his findings are either not mentioned or – when they are – somebody else is taking credit for them.

But I want to make clear that the following overviews must not be taken as substitutes for the real books. Their sole purpose is to show that it is worth while carefully reading the whole books. Only there you can find all the details, arguments and sources which are necessary for a critical in depth-discussion of the topics. Any criticism which is only based on an overview disqualifies itself as frivolous – like the many many cases in the past decades.

The overviews were written by C. J. Ransom, Ph. D., a physicist who received his doctorate in plasma physics at The University of Texas in Austin, Texas, and conducted research and technical computing in the Aerospace industry for over 30 years. Dr. Ransom is a life member of the American Physical Society (APS) and belongs to the Institute of Electrical and Electronics Engineers (IEEE). Furthermore he was the executive director of Cosmos & Chronos, a non-profit research organization that derived from a group created by Professor H. H. Hess while he was Chairman of the Geology Department at Princeton University, and authored the book "The Age of Velikovsky" (Dell Publishing Company, 1978).

Overview of *Worlds in Collision*

In 1950, Macmillan published Dr. Immanuel Velikovsky's book *Worlds in Collision*. Much of the academic community complained about what they assumed he said and forced Macmillan to stop publishing the book. Doubleday, who was not influenced by the irrational equivalent of book burning, published the book after Macmillan stopped.

What could someone say that was not about religion that could produce such outlandish human behavior? Part of the problem was that some people incorrectly assumed the book was written to support religion, but that is a side issue. Much of the scientific community was opposed to the book, and stopped Macmillan from publishing it, because *Worlds in Collision* challenged their basic assumption of uniformity. The book also challenged the basic assumption of a neutral Universe.

Uniformity is the principle that all geological features on Earth can be explained by processes now seen occurring on Earth. No global catastrophes had occurred after the creation of Earth X billion years ago, and certainly no catastrophes were caused by bodies external to Earth. Astronomers had their version of uniformity. They assumed that all planets were formed in their present orbits X billion years ago and no changes occurred after that. Velikovsky said both of those assumptions about uniformity were wrong. The reaction was equivalent to what would happen if he had said that the Christian, Hebrew, Hindu, and Muslim religions are wrong.

In the preface to *Worlds in Collision*, Velikovsky said, "The reader is not asked to accept a theory without question." He had assumed, however, that people would not reject it without a rational analysis. He also assumed that as people did rationally investigate evidence of past unusual events and discovered that Velikovsky was often right, they would give him some credit for some of his conclusions. He was disappointed to discover that neither happened.

Today, the scientific community agrees that neither the geologists' or astronomers' 1950 version of uniformity is true. Much of what Velikovsky said has been republished, if not independently rediscovered, by scientists since 1950. Some of the reworked ideas were by scientists who originally claimed that Velikovsky was wrong. Unfortunately, most people only remember that scientists in 1950 told the public that Velikovsky was wrong. They do not remember what he said, why he said it, or that he was often right.

The following is an overview of some of the ideas in *Worlds in Collision* and how he arrived at those ideas. I do not cover his model in detail about the recent history of the Solar System because new data are available. Models are modified as new data become available. It would not be proper to present details of his model and include modern data as if he had, or could have, used that information.

Trying to present his detailed model also has the problem that I may not include enough important pieces of information and, hence, may give the wrong idea because of my selection. In addition, *Worlds in Collision* does not cover all major events BCE. He said in the preface to the book, "*Worlds in Collision* comprises only the last two acts of the cosmic drama." He planned to provide a more detailed model including additional events. He did not have time to complete that analysis and did not include much about it in *Worlds in Collision*. Therefore, the following is a description of the big picture of his ideas in *Worlds in Collision* and a small amount of detail about what led to the big picture.

The following contains a lot of data considered by Velikovsky that needs to be part of any detailed model. I present some of the information he compiled in *Worlds in Collision* for that reason, and not to try to use it to support his detailed model. An overview naturally cannot contain all the necessary information to prove a point. Velikovsky also had extensive documentation on each page. In the overview, I will not refer to his sources.

The basic premise of *Worlds in Collision* is that mythology is not fiction derived for entertainment. Velikovsky said that ancient cultures saw images in the sky that they interpreted as activities of the gods. Years after the events, later people wrote the stories handed down by the culture. Those stories were called mythology. Velikovsky said that ancient myths contain data. He said that detailed comparison of mythology can be used to help determine what events ancient cultures saw in the sky.

To do that, Velikovsky looked for similar features of myths from different cultures. He used comparative mythology to try to identify archetypes that may have had local embellishments but that had a number of identical basic characteristics. Many other people used comparative mythology, but generally they were interested in what appeared to be psychological causes of the similarities. Velikovsky reasoned that the ancient cultures were describing a physical reason for creation of the myths.

In order to determine what the ancients said caused the events, Velikovsky needed to determine the basic mutual components of mythology. He said, "We shall check one people against another, one inscription against another, epics against charts, geology against legends, until we are able to extract the historical facts." (*Worlds in Collision*, Preface 1950)

He cautioned that often it is impossible to be certain whether a particular tradition refers to which catastrophic event. He also noted that the difficulty is compounded because some traditions may have elements mixed with other traditions. With those cautions, he concluded that (1) physical upheavals occurred during the time humans were around to see the events, (2) physical bodies outside Earth caused the events, (3) the bodies could be identified, and (4), electromagnetism plays an important role in the Solar System and Universe.

In the first two chapters, Velikovsky reviewed some of the 1950 scientific beliefs and problems confronting theories about the Solar System. He described the immense Universe and how important man considered himself and how much we thought we knew at that time. He then described how small we are, how little we know. Unknowns he mentioned included the origin of oil, the origin and disappearance of the ice sheet in much of Europe and North America, why there is a recorded history of only a few thousand years, and why the Bronze Age preceded the Iron Age.

Velikovsky noted that currently there is celestial harmony. People today expect the sun to rise as always, the moon to be there when it is supposed to be, the axis to point to the North Star. The current assumption is that it was always that way and always will be that way.

He also described several known characteristics about the Solar System at the time (1950). A few turned out not to be true, but that is another article. He pointed out that the Solar System bodies did not have characteristics that one could easily explain by assuming creation of the bodies from a nebula and the bodies being unaffected since the origin. If they did have a similar origin and characteristics, he said they obviously changed. For example, he noted the axes of rotation are not tilted the same direction for all planets, the moons do not all spin about their primary planets in the same direction, and the number of moons varies. If everything started from the smooth development from a nebula, something must have made the odd characteristics of some planets and some moons. He wondered why we insist that all of the changes occurred "once upon a time, long, long ago." Velikovsky then described some of the major ideas concerning the origin of the Solar System, pros and cons of the ideas. None of those ideas considered any recent changes.

Velikovsky then described some of the basic data about comets and ideas believed in 1950 about the origin of comets. He presented that information because he later demonstrated that the 1950 ideas conflicted with what the ancients appeared to have observed.

He also described a number of concepts about Earth and its characteristics. He noted that many of the characteristics of Earth appear to have been caused by major events, but that the theories about those characteristics assume uniformity.

Velikovsky presented that information as a background of the 1950 understanding about the Solar System and Earth. He then discussed reasons that the model that was accepted at the time was not necessarily the most accurate model for the history of the Solar System, and especially Earth.

Velikovsky noted that a common feature of ancient cultures is the concept of the world age. World ages ended in violent events. Different cultures had different numbers of world ages. The number of world ages depended on factors such as the influence of a particular catastrophe on a group or the way a group decided what constituted a world age. He gave a number of examples of different cultures and their descriptions of world ages.

Writers in ancient Eturia said there were seven ages. Hesiod, and early Greek writer, said there were four ages. The Hindu book *Bhagavata Purana* described four previous ages and said we are now in the fifth age. Some Chinese said there were ten ages from the beginning until Confucius. Some texts in Mexico discuss four ages. Some rabbinical sources say that we are in the seventh age. Those sources talk about the seven "earths" and seven "heavens." There are numbers of other examples of ancient cultures describing world ages. Sometimes the ages are called "World Cycles."

Often some cultures mention a new sun at the beginning of the new age. The suns mark the epochs associated with various catastrophes. Sometimes the ages are named for the predominant type of catastrophe associated with the age. All types of catastrophe were involved in each age, but one was more noticeable for that age. The suns were named for deluge, hurricane, earthquake and fire. Areas of Mexico describe five suns. A Roman author also wrote of five suns. Some sources in Mexico mention seven suns. Some Buddhist traditions refer to seven suns. Aborigines of British North Borneo even in 1950 said that we are presently in the seventh sun.

Velikovsky asked, "Did the reason for the substitution of the word 'sun' for 'epoch' by the peoples of both hemispheres lie in the changed appearance of the luminary and in its changed path across the sky in each world age"?

(Comment: Velikovsky used data from the Old Testament that correlated well with data from other sources. Some people mistook this as an attempt to support religion. Some scientists complained that he was trying to support religion and some religious people complained that he was trying to explain away religion. This created a number of irrational arguments before encountering the scientific aspect. Velikovsky was not trying to support religion. It just happens that a number of Biblical stories match information from around the world. Currently, there is also reason to believe that the timing of the events mentioned in the Bible are not when the events really occurred.)

Many ancient stories describe a time when things turned red. At that time, some things were renamed red sea, red-mountain or red whatever. Some interpreted red water as having turned to blood. Velikovsky said that this is often the first event leading to a great catastrophe. It is also what could be expected by Earth entering the atmosphere of a comet tail containing fine red dust particles.

The observers did not know about comets, so they designated the red as the blood of various gods. Velikovsky gave references to stories from places such as Egypt, Finland, and Babylonia.

As Earth entered further into the atmosphere of an external body, the particles became larger. Small dust became stones.

Velikovsky noted that two verses before the Biblical description of the sun standing still the story relates that stones fell from heaven. Joshua 10:11 (King James Version) says "And it came to pass, as they fled from before Israel, and were in the going down to Bethhoron, that the LORD cast down great stones from heaven upon them unto Azekah, and they died: they were more which died with hailstones than they whom the children of Israel slew with the sword." Velikovsky said that the writers of Joshua were unaware of any connection between rocks falling from the sky and a possible disruption in the apparent movement of the sun. Many ancient cultures describe a different age when red-hot stones fell from the sky.

Ancient stories connected stones falling from the sky, earthquakes, whirlwinds and a disturbance in the movement of Earth. This connection was not accidental. The four phenomena could be expected if a large object external to Earth had a close encounter with Earth.

In the other hemisphere, the Mexican *Annals of Cuauhtitlan*, also known as *Codex Chimalpopoca*, described a cosmic catastrophe in which the night was prolonged. This would be expected if a longer day occurred in the Middle East. Many ancient cultures describe a different age when the spans of days did not appear to be what was normal.

Velikovsky concluded that the event described in those stories was a smaller event. One or more greater events took place earlier.

The pre-Colombian traditions of Central America said that another major catastrophe occurred fifty-two years before the catastrophe involving the long night. This fit the time frame of another catastrophe in the Middle East. This is the place in *Worlds in Collision* where Velikovsky mentions the connection to his work *Ages in Chaos*, which was written earlier, but published later. *Ages in Chaos* identifies oddities in the chronology of Egyptian history. If those problems can be fixed the way Velikovsky suggested, then a major catastrophe described by the Hebrews and one described by the Egyptians coincide. This one to one correlation of revised Egyptian history and Hebrew history makes some of the events of the catastrophes fit closely, but it is not a requirement for the identification of the details of the catastrophes. (All of the timing of the stories may have problems because of the time of the writing of the stories, and, hence, the time period associated with the stories, was not when the events occurred.)

Ages in Chaos can be right or wrong and not affect the overall thesis of *Worlds in Collision*. People interested in historical chronology may want to read *Ages in Chaos* to see the complex problems in establishing an accurate Egyptian chronology.

After the stones fell, various cultures described a time when a petroleum based product fell from the sky. Sometimes the hot rocks were mingled with the fire from the petroleum products. Again, people in the Middle East and the Americas had similar stories. The flammable substance that fell from the sky was used in fires in worship places and in homes. Use of the substance decreased as the part that stayed above ground was consumed.

As Earth entered even further into the atmosphere of the external body, the smog created what became known as the time of darkness. The length of time of the darkness varies among different cultures. This could be expected if your main method of telling time was a sun dial and you could not see your hands in front of your face. People in the Old World and the New World reported a similar event.

If the event were caused only by smog, you would expect the same story through the world. However, some people talk about an extended night followed by an extended day. Iranians talk about a time when there was a light in length of about three days followed by a night of about three days. At that time, people in China said the sun did not set for about ten days. Velikovsky attributed the change in the appearance of the sun possibly to a tilt of Earth's axis or maybe even a slowing of the rate of rotation.

As the external body neared its close approach to Earth, most of the world endured earthquakes. The effects were such that the world also had numerous exceptionally large hurricanes and strong winds. Ancient Mayan sources tell of a time when a terrible hurricane swept the earth and broke up and carried away the towns and forests. Hurakan is the ancient Mayan god of wind and storm. The word hurricane is derived from that name. Buddhist texts say that the wind turned the ground upside down and threw it into the sky. In a Japanese cosmological myth, the sun goddess hid herself in a cave in fear of the storm god. In Japan hurricanes and earthquakes are common, but they do not affect the day-night succession and do not make permanent changes in the appearance of the sky.

The close approach of the external body also brought tidal waves. People throughout the world describe a time when the sun's movements were disrupted and great tidal waves appeared. Velikovsky mentions stories from diverse sources such as China, the Choctaw, Peru, and the Midrashim.

To the frightened observers throughout the world, the images in the sky appeared to be battles among the gods. The time also produced extensive lightning storms. Large lightning bolts appeared to have struck between the external body and Earth, hence the lightning bolts of the gods. The tail of the receding external body then appeared as a serpent or dragon-like image. The fight appeared to be between the big glowing orb in front of or attached to the snake-like tail. The events appeared to be a fight between the orb god and the snake god. A god who threw his thunderbolt at the world and is pictured with lightning in his hand has various names. He was Zeus of the Greeks, Odin of the Icelanders, Ukko of the Finns, Perun of the Russian pagans, Wotan (Woden) of the Germans, Mazda of the Persians, Marduk of the Babylonians, Shiva of the Hindus.

This event is described in numerous stories throughout the world. Velikovsky said, "It is difficult to find a people or tribe on the earth that does not have the same motif at the very focus of its religious beliefs." (Part I, Ch. 3, »The Battle in the Sky«) Some people who saw the images from the Middle East called the event the fight between Zeus and Typhon. The snake lost the battle and wound up at the bottom of the sea. A name for Typhon used by the Greeks was Pallas. The Egyptian equivalent was Seth. If Velikovsky's premise in *Ages in Chaos* is correct, the Egyptian event correlates in description with disasters associated with the Exodus.

The series of events is a set of exceptional disasters followed by the consequences of the disaster. For example, if the water is polluted, fish die and make the area smelly. Velikovsky correlates the natural disasters and the consequences of those disasters with the ten plagues mentioned in Exodus. (Comment: Several authors since then have tried to explain the ten plagues in the same manner. The reason Velikovksy's series is possibly more reasonable is that he has an explanation for the first disaster that started the chain reaction and the one that caused the most destruction.)

Earthquakes and volcanoes are known to produce sounds. The *epic of Gilgamesh* talks about a time when the firmament roared and the earth responded. In Hesiod it says that the earth groaned when Zeus hit Typhon with his bolts. Homer said the earth ring and the heaven pealed with a trumpet sound. A writing attributed to Confucius said the Emperor Yaou (Yau) (Yahoo) was first called Shu King. The name Yahou was given to Shu King after the catastrophe and was possibly inspired by the sounds from the earth at the time. The same sound was heard in the Western Hemisphere. Native Americans relate a story about when the heavens were close to the earth and all mankind lifted the sky by shouting "Yahu." Velikovsky noted that the name Yahweh is preserved in shorter forms such as Yahou and Yo for the name of the Deity in the Bible. In Mexico, Yao or Yaotl is the god of war.

Velikovsky continued with a detailed discussion about the Chinese Emperor Yahou. He did that to demonstrate the nearly identical traditions of the Chinese during the time of Yahou and Jewish people during the Exodus. These traditions include the sun disappearing for a number of days, the land being overrun with vermin, there were gigantic tidal waves, the world burned, and a new calendar was established because the apparent views of the heavens and the seasons were different.

Velikovsky then went into considerable detail about other sources that described the changes in the movements of the earth and the other characteristics shared by the Chinese and the Hebrews. These other characteristics included the time of darkness, which the Hebrews called "the Shadow of Death", an edible hydrocarbon based material in the smog, which the Hebrews called "Manna" and others called such names as "Ambrosia".

Velikovsky presented evidence that many cultures describe two different times when there were catastrophic floods. The Greek called the most devastating one the flood of Deucalion. In that flood, supposedly only two people remained alive to repopulate the world. Velikovsky said, "This last detail must not be taken more literally than similar statements found in descriptions of great catastrophes all around the world; for example, two daughters of Lot, who hid with him in a cave after the catastrophe of Sodom and Gomorrah, believed that they and their father were the only survivors in the land." (p. 148)

The other flood in Greek was that of Ogyges. Velikovsky provided support from various sources that that flood was associated with the events in the days of Joshua, and the flood of Deucalion with the days of the Exodus.

The span between the catastrophes of the Exodus and those at the time of Joshua was about fifty-two years. Velikovsky noted that numerous other cultures thought catastrophes occurred about fifty-two years apart. The natives of pre-Columbian Mexico expected a new catastrophe at the end of every fifty-two year period. Both the Mayas and the Aztecs associated this period with Venus.

Ancient Mexican records give information about a snakelike body adorned with feathers that transformed itself into a great star. The feathered arrangement also represented flames of fire. This great star first appeared in the east. The snakelike body was called Quetzal-cohuatl. The great star, or planet, retained the name Quetzal-cohuatl. That is also the name for Venus.

After the sun did not appear for four days, the seasons were disarranged. Many people died from famine and pestilence. There was also a deluge.

Chinese chronicles said that a brilliant star appeared in the days of Yahu [Yahou]. A Samaritan chronicle said that a new star was born in the east. Varro wrote about the Roman people and a time when the brilliant star Venus changed its color, size, form, and course. He said it never happened before or since.

The birth or the transformation of a legendary person into the morning star is a widespread motif. Examples are Phaethon, Quetzal-cohutal, Istehar, Atymnios. The Tahitians, the Buriats, Kirghiz, and Yakuts of Siberia, and the Eskimos of North America have the same tradition. Apollonius Rhodius in *The Argonautica* mentions a time when not all the planets were yet in the heavens.

Modern writers believe that ancient writers were confused when they referred to Venus as having comet-like characteristics. They said Venus had a tail and called it the star that smoked. Venus was also said to have a beard. Ancient writers also talked about a time when Venus appeared to grow horns. Again, these are world wide motifs. There was no confusion on the part of ancient writers.

Also widespread was the worship of the Morning Star. Velikovsky noted that Isaiah 9:2 was translated as "The people who walked in darkness have seen a great light; Those who dwelt in the land of the shadow of death, Upon them a light has shined." He noted that the original Hebrew said the light was referred to as the light of Noga. He said the Noga was the Hebrew name for Venus and it was an omission not to include it in the translation.

The various gods associated with Venus were also associated with a serpent or dragon. The planet was also associated with a solar disk at its back. Many ancient cultures held festivals for the planet Venus in April. Some cultures made human sacrifices to the Morning Star. Sacred cows and bull worship became prevalent about the same time. The celestial cow damaged the earth. The morning star was given name equivalents of Satan, such as Lucifer and Seth.

The synodical year of Venus was used as a religious calendar by many cultures. The religious significance was preserved into the Middle-Ages. This was done from Egypt to the Americas. The movements of Venus were carefully watched because the ancients wanted to be sure that Venus was not a threat to Earth.

Although less destructive than Venus, many diverse ancient cultures said Mars later created havoc on the earth. Mars appeared to have several encounters with Earth. These events are described in detail by such writers as Amos and Isaiah. Other cultures mentioned the activities of Mars and worshiped Mars.

Romans acquired most of their mythology from the Greeks. The roman culture was younger than the Greek. Velikovsky suggested that since the Romans did not have all of the previous mythological baggage of older cultures, the Romans gave the more recent Mars events a more important place in their mythology. Mars, who had the counterpart Ares in Greece, was the lord of war, the national god of the Romans, and supposedly the founder of Rome. A Roman month, March, was dedicated to Mars.

Mars was also associated with disruptions on Earth. Velikovsky said the events described are nearly identical to those described in the Bible concerning the time of Hezekiah and Sennacherib. The Babylonians called Mars "Nergal". They considered Nergal to be the most violent among the gods. Mars was also called the unpredictable planet. Chinese records called Mars "Ying-Huo", or the fire planet. The Aztec god Huitzilopochtli (Mars) was greatly feared in warfare.

Velikovsky noted that Venus was feared first and then Mars became important as a deity. He suggested that this timing may have been because of a Venus-Mars encounter. He further noted that if Venus and Mars had near encounters, the events would have been seen from Earth. He said that the *Iliad* and *Odyssey* appeared to contain descriptions of various Venus-Earth, Venus-Mars and Mars-Earth encounters. He said that the Greeks having Venus as the major god and the later Romans having Mars as the major god was the same sequence as in the new world. The Toltecs had Venus as a major god and the later Aztecs revered Huitzilopochtli (Vitchilupuchtli) as the major god.

Chinese records mention a time when two orbs battled in the sky. The two battling stars were Venus and Mars. The *Soochow Astronomical Chart* said that some of the ancients said the planets went off course and Venus ran far from the zodiac and attacked the "Wolf-Star." Hindu records in *Surya-Siddhanta* mention encounters between planets. They said Venus generally wins. The early Hindus knew that the earth was a sphere and was one of the planets and knew that above and beneath were only relative terms, so they were advanced to make useful observations. The *Bundahis* also describes encounters of planets in the sky that created chaos in the cosmos.

Another reference to Mars was the Sword-God. Images of Mars were pictured with a sword. Comets that appeared as swords were related to the planet Mars. The Scythians worshipped Mars and their image of him was a scimitar of iron.

The Martian atmosphere took on various shapes in addition to a sword. The shape appeared as a lion, jackal, dog, pig, and fish. One form that was noted by many ancient cultures was that of a wolf or jackal. Cultures as widespread as Egypt and China called Mars the wolf star.

Velikovsky noted that many ancient myths refer to the gods throwing large lightning bolts at the earth and other gods. He further noted that lightning is often seen being discharged between two clouds or a cloud and the ground. He then said, "But if for some reason the charge of the ionosphere, the electrified layer of the upper atmosphere, should be sufficiently increased, a discharge would occur between the upper atmosphere and the ground, and a thunderbolt would crash from a cloudless sky." (Part II, Ch. 4, »Sword-Time, Wolf-Time«) (Comment: This is now known to be correct.)

Thunderbolts could also occur between two close bodies with oppositely charged atmospheres.

Ancient stories show gods holding thunderbolts and throwing them at other gods and the earth. The Talmudic and Midrashic sources indicate the army of Sennacherib was destroyed by a blast. This blast was attributed in some sections of the Scriptures to an angel. Velikovsky considered that his detailed analysis demonstrated that the angel was Mars. He further concluded that the Archangels were in some cases planets. For example, he noted that Gabriel was connected with the foundation of Rome; therefore, Gabriel was Mars.

The previous reviews a small amount of the data Velikovsky amassed that led him to conclude that Venus and Mars were important deities of many ancient cultures and that those planets created havoc on the Earth and on each other. The data need to be explained by any model trying to reconstruct the events before the Solar System became stable. In *Worlds in Collision*, Velikovsky provided some suggested timing for various events, but what is most important is the identification of a different ancient sky. After demonstrating that the ancient sky was different, he delved into the psychological aspects of why the events were suppressed into "mythology."

Velikovsky coined the term "collective amnesia," but many people have used it since. He described the process where a terrifying event in childhood can be forgotten. The memory is blotted by the consciousness and sent to the unconscious area of the mind. The memory, however, continues to live and express itself in bizarre forms of fear. He applied the concept to the collective group of mankind. The literally earth shattering events were so terrifying that people as a group suppressed the events and committed them to mythology. People still try to relive the events through war and other irrational acts.

Velikovsky then added to the enormous data set by describing various physical anomalies that were not written as myths and that indicate something unusual happened in the past. For example, he noted that according to Seneca, the Great Bear was the polar constellation until after a cosmic upheaval shifted it to the Little Bear. The *Jaiminiya-Upanisad-Brahmana* made a similar statement. Early Egyptians also said that the Great Bear never set. After the change, many ancient cultures believed that if the polar star moved from its place, the Earth would be destroyed in a great conflagration.

Another anomaly is the measured length of the longest and shortest days. Chinese measurements show the longest day duration does not represent the various geographical latitudes of their observations. Eighth century Babylonian astronomical tablets provide exact data, but the data do not match modern observations.

Temples of the ancient world faced the sunrise. Many had indicators that marked the extremes of the equinoxes. Investigators have found that there were deliberate changes in the orientation of the foundations of some older temples. Generally, only the more recently built temples faced east, and temples built before the seventh century BCE did not face east. The same non-east orientation was found in a number of archaic foundations.

A sundial in Egypt that dated about 850 BCE to 720 BCE does not work for where it was found. Velikovsky suggested that analysis of that shadow clock may be useful in determining the inclination of the pole to the ecliptic and the latitude of Egypt at that location during the time before the current alignment of the Earth.

A water clock found in Egypt also does not provide the expected readings for the current alignment of the Earth. For the winter solstice, the clock is off by fifty-two minutes. The summer solstice and the equinoxes are also incorrect. The differences are consistent.

About 1939 a startling find was made at Point Hope in Alaska, on the shores of Bering Strait. Rainey and colleagues found an ancient city that had about eight hundred houses. The population was estimated to be larger than the then population of Fairbanks. Point Hope is about 130 miles inside the Arctic Circle. This site, the largest and oldest site yet found in the Arctic, led to a reconsideration of all known theories regarding the origins of the Eskimo or Inuit peoples. Velikovsky used information of this type and such as the temple orientations, water clocks, and observed pole star locations to conclude that the earth's axis is not what it was in pre-historical times.

He also discussed numerous inconsistencies in calendars from ancient times. Observations made by various peoples during the last 2500 years are extremely accurate. There is no known technology advancement that would have made those observations more accurate than those before that time, yet previous calendars were no longer useful. Some of the post 500 BCE calendars are based on observations of Venus. It seemed important to track the movements of Venus as well as the Sun. This would be expected if there had been changes and if Venus had created havoc on the Earth.

At the end of *Worlds in Collision*, Velikovsky used a little space to describe what he expected would eventually be found on the Moon, Mars and Venus as a result of the recent history of those planets. That information is covered in detail in my *High-Level Velikovsky* [published in *The Truth Behind the Torment*].

In the conclusion, Velikovsky indicated that combined Solar System changes and Earth changes are reflected in what we call mythology and that we can determine the changes from comparative mythology. He said, "The solar system is not a structure that has remained unchanged for billions of years; displacement of members of the system occurred in historical times. Nor is there justification for the excuse that man cannot know or find out how this system came into being because he was not there when it was arranged in its present pattern."

Overview of *Earth in Upheaval*

Opponents of Velikovsky gave numerous excuses, not reasons, for not investigating his concepts. The primary excuses were basically invocations of two accepted assumptions of 1950. These assumptions were (1) no changes had occurred in the Solar System since its origin and (2) no catastrophes had occurred on Earth after its origin. A secondary excuse was that Velikovsky used observations of the ancients and the assumption was (3) that the ancients must have been poor observers. Velikovsky wrote *Earth in Upheaval* in response to item (2) and (3). He used geological data but no documents from the ancient world to demonstrate that the geological record indicated that catastrophes had occurred on Earth, no matter how old Earth was. Another secondary excuse was (4) Velikovsky used old references. Recent geologists have used some of the same references and new data to form the same conclusions. Properly collected scientific data do not expire.

In the overview, I will describe types of data Velikovsky used and not go into very much detail. There are two reasons for that approach. First, geologists now agree that Velikovsky was right, although they will still not say it in those words. Second, there is now newer abundant data indicating that Velikovsky was right so there is little reason to cover the old data in detail.

Not all geologists at the time refused to work with Velikovsky. Several geologists read portions of *Earth in Upheaval* and provided suggestions about additional references. These included Professor Waldo S. Glock, Chairman of the Department of Geology at Macalester College, a recognized authority in dendrochronology, Dr. T. E. Nikulins, geologist in Caracas, Venezuela, and Professor George McCready Price, a geologist in California. People in other fields were also helpful to Velikovsky in developing *Earth in Upheaval*. These included Dr. Albert Einstein, Professor Lloyd Motz of the Department of Astronomy at Columbia, Dr. H. Manley of the Imperial College, London, Professor P. L. Mercanton of the University of Lausanne, Professor E. Thellier of Observatoire Géophysique at the University of Paris, and Professor Richardson of the Illinois Institute of Technology.

Velikovsky described a number of geological sites that indicated sudden changes during the Ice Age era. Information was taken from descriptions by geologists who even used the word "catastrophe". Sites were in Alaska in the US and Siberia in Russia. Velikovsky quoted the German scientist G. A. Erman who went to the Liakhov and the New Siberian islands. Erman found areas full of bones of elephants, rhinoceroses and buffaloes. He said, "In New Siberia [Island], on the declivities facing the south, lie hills 250 or 300 feet high, formed of driftwood, the ancient origin of which, as well as of the fossil wood in the tundras, anterior to the history of the Earth in its present state, strikes at once even the most uneducated hunter". Also, "On the summit of the hills they [the trunks of trees] lie flung upon one another in the wildest disorder, forced upright in spite of gravitation, and with their tops broken off or crushed, as if they had been thrown with great violence from the south on a bank, and there heaped up". (Ch. 1, »The Ivory Islands«)

Erratic boulders are rocks that differ from the formations on which they lie. Some are transported by glaciers. Some are found in the opposite direction from their origin from the direction of movement of glaciers. Those generally are thought to have been moved by tidal waves. There are loose rocks lying on the Jura Mountains with mineral compositions depicting an Alpine origin. One of those erratic boulders is over 10,000 cubic feet. There are erratic boulders in many places throughout the world. Many are extremely large.

There are many places where sea strata and fresh water strata are stacked. Georges Cuvier said, "These repeated irruptions and retreats of the sea have neither all been slow nor gradual; on the contrary, most of the catastrophes which have occasioned them have been sudden; and this is especially easy to be proven with regard to the last of these catastrophes, that which, by a twofold motion, has inundated, and afterwards laid dry, our present continents, or at least a part of the land which forms them at the present day". (Ch. 2, »Sea and Land Changed Places«)

A cave in Kirkdale in Yorkshire is eighty feet above the valley. Under a covering of stalagmites, there are teeth and bones of elephants, rhinoceroses, hippopotami, horses, deer, tigers, bears, wolves, hyenas, foxes, hares, rabbits, as well as ravens, pigeons, larks, snipe, and ducks. Investigators noted that the presence of the tropical animals in northern Europe could not be explained by periodic migrations.

One of the oldest strata with signs of extinct life in it is called the Old Red Sandstone in Scotland. Many remains exhibit the marks of violent death. The figures are contorted, contracted, and curved. In many cases the tail is bent around to the head. The fins are spread out in full as in fish that die in convulsions. A similar picture is found throughout the world. W. Buckland wrote about fish deposits in the Harz Mountains in Germany. He said they were buried before putrefaction started. Velikovsky concluded, "In cataclysms of early ages fishes died in agony; and the sand and the gravel of the upthrust sea bottom covered the aquatic graveyards". (Ch. 2, »The Aquatic Graveyards«)

Velikovsky described the origin of the idea of uniformity. It largely depends on the idea that everything seen in the geological record can be explained by processes now seen occurring. This provided a method to demonstrate the extreme age of the earth countering the religious idea of a young earth. Even earthquakes, hurricanes, volcanoes and tidal waves, however, cannot explain many of the features found in the fossil record. Velikovsky presented many of those records. He also noted that even Darwin saw evidence of catastrophes that he could not explain.

Velikovsky said that catastrophes were the friend of evolution. They would provide an explanation for mutations and changed environments conducive for evolution. He even coined the term "Cataclysmic Evolution". This idea is now widely accepted, however, without mention of Velikovsky.

Velikovsky discussed corals found in polar-regions and other evidence that warm and cold places had drastically changed climate. At the time, geologists rejected the idea of continental drift, later described as plate tectonics. The rejection in part was because there was no mechanism for movement of the continents. Velikovsky said that catastrophes caused by agents external to Earth may be partially responsible for the energy needed to move the continents.

Moving continents, however, do not move fast enough to account for the recent ice ages. W. B. Wright of the Geological Survey in Great Britain said that the best way to explain the ice ages was a shift of the direction of the earth's axis. This suggestion was discarded because of a mechanism to cause the shift. Velikovsky's concept could provide the mechanism.

Coal resulted from catastrophes. Slow accumulations and covering of debris cannot explain the contents or sizes of coal beds. Velikovsky said, "Forests burned, a hurricane uprooted them, and a tidal wave or succession of tidal waves coming from the sea fell upon the charred and splintered trees and swept them into great heap, tossed by billows, and covered them with marine sand, pebbles and shells, and weeds and fishes; another tide deposited on top of the sand more carbonized logs, threw them in other heaps, and again covered them with marine sediment. The heated ground metamorphosed the charred wood to coal, and if the wood or the ground where it was buried was drenched in a bituminous outpouring, bituminous coal was formed. Wet leaves sometimes survived the forest fires and, swept into the same heaps of logs and sand, left their design on the coal. Thus it is that seams of coal are covered with marine sediment; for that reason also a seam may bifurcate and have marine deposits between its branches". (Ch. 13, »Coal«)

(Comment: Just before the publication of *Earth in Upheaval* in 1955, Velikovsky found an extensive work by Nilsson published a short time earlier. Nilsson was a professor of botany at Lund University, Sweden, when he published the two volume work entitled *Synthetische Artbildung*. He discussed results of various botanical studies of certain coal deposits. In Nilsson's opinion, debris trapped in the coal was deposited by onrushing water from all parts of the world, but mostly from the coasts of the equatorial belt of the Pacific and Indian Oceans. Wilfrid Francis, author of *Coal, Its Formation and Composition*, in the second edition of his book [the first edition was printed before *Earth in Upheaval*] agreed that Nilsson's and Velikovsky's explanation for the origin of coal is the best explanation.)

Velikovsky provided numerous examples of geological data that supports the concept of major catastrophes larger than would be expected from commonly observed agents causing what are thought to be catastrophes today. The information is very interesting and clearly presented, but it is no longer needed to convince geologists that Earth was involved in many catastrophes caused by agents external to Earth. The only difference now between geologists, astronomers and Velikovsky is when the last catastrophe occurred. The accepted scientific community still prefers to consider all catastrophes as having occurred "Once upon a time, long, long ago." That opinion should change as more and more comparative mythologists become involved in the investigation.

Overview of *Ages in Chaos – From the Exodus to King Akhnaton*

In 1952, Doubleday published Velikovsky's book *Ages in Chaos*. That book is unlike his *Worlds in Collision* and *Earth in Upheaval*. Those books dealt with the recent physical catastrophic history of the Earth and the extended geological history of the Earth. *Ages in Chaos* is a book largely about Egyptian history. Velikovsky was working on that book when he noticed possible similarities between ancient texts from Egyptians and Hebrew sources. From 1940 on, he worked on both *Ages in Chaos* and *Worlds in Collision*, the latter published in 1950.

By comparing ancient sources, Velikovsky noticed that they seemed to describe similar events, but the events were hundreds of years apart based on conventional Egyptian history. This led him to consider the possibility that the chronology of one of the nations was incorrect. After detailed analysis of numerous historical documents, he concluded that there are some problems with the Egyptian chronology. In the foreword, he noted that historians may have greater psychological difficulties in this revision of their views than astronomers had in considering the possibility of catastrophes on the earth caused by external agents. He also said, "Is it not the case that at first a new idea is regarded as not true, and later, when accepted, as not being new?" (*Ages in Chaos*, Foreword) He was correct about both.

Although Velikovsky took personal responsibility for his books, he did not work in a vacuum. Velikovsky always consulted experts in various fields before publishing a work. *Ages in Chaos* was no exception. For example, he was assisted by Dr. Robert H. Pfeiffer who was the director of the Harvard excavation at Nuzi, curator of the Semitic Museum at Harvard, and professor of ancient history at Brown University. Pfeiffer kept an open mind about the issue because he believed that only objective and free discussion could decide the matter. Other scholars also provided Velikovsky with leads to important information.

There are two major components to Velikovsky's chronological revisionism. The first relates to abundant data indicating that there are problems with the accepted chronology and the second is his proposed revision. *Peoples of the Sea* contains a supplement called »Astronomy and Chronology«. That section demonstrates that the astronomical anchors for Egyptian chronology are considerably less substantial than advertised.

Ages in Chaos starts a search for a link between ancient Israelite and Egyptian histories. Israelite history describes a time when they were in bondage in Egypt and left during catastrophic events. Historians disagree about the exact date of the Exodus, but agree the event occurred during what is called the New Kingdom of Egypt. Egyptian history, however, supposedly does not mention the Israelites in Egypt during that period or even the catastrophic events.

Identifying the time of the Exodus was important in order to compare to the Egyptian history. Traditionally, the Exodus occurred about 480 years before Solomon built the

Temple. This would make the Exodus around 1440 BCE. The Egyptian Pharaohs were in complete control of Canaan at that time, which was during the New Kingdom time frame. Yet, Joshua did not find a strong Egyptian hold there after the Exodus.

Velikovsky reviewed several theories of the time trying to connect the Exodus with Egyptian history. Three of the theories were, respectively, based on expulsion of the Hyksos from Egypt, invasion of the Habiru, and destruction of Israel in the days of Merneptah. Velikovsky noted that "It is hopeless to try to reconcile the irreconcilable." (Ch. 1, »What is the Historical Time of the Exodus?«) He gave a number of problems with each theory. Historians considered the date of Israel's settlement of the Palestine area to be in the time of about 1500 – 1100 BCE. Under the accepted chronology, the New Kingdom of Egypt was too strong for the Exodus to have occurred and they do not mention Israelites living in Egypt. Similarly, the Israelites do not mention Egypt in the Books of Joshua and Judges.

The Exodus describes a series of catastrophes, or plagues, occurring at the time. Believing that there should be references to similar events in Egyptian history, Velikovsky reviewed documents from ancient Egypt. He discovered that there was a document called *Admonitions of an Egyptian Sage* in the Museum of Leiden in the Netherlands. The sage was named Ipuwer and Velikovsky referred to the document as the *Papyrus Ipuwer*. The document was translated in 1909 by Alan H. Gardiner.

The *Papyrus Ipuwer* described the social system as becoming disorganized, there was violence throughout the land, invaders preyed on the defenseless population, the rich lost their possessions and the whole country was in distress. Gardiner claimed that events described were sociological in nature.

Velikovsky correlated descriptions of the plagues and the words from the *Papyrus Ipuwer* and demonstrated that they could easily be describing the same set of events. Some examples of comparison follow:

EXODUS 7:21   ... there was blood throughout all the land of Egypt.

PAPYRUS 2:5-6   Plague is throughout the land. Blood is everywhere.

EXODUS 7:24   And all the Egyptians digged round about the river for water to drink; for they could not drink of the water of the river.

PAPYRUS 2:10   Men shrink from tasting – human beings, and thirst after water.

EXODUS 9:25   ... and the hail smote every herb of the field, and brake every tree of the field.

(Also PSALMS 105:33   He smote [with hail] their vines also and their fig trees; and brake the trees of their coasts.)

PAPYRUS 4:14   Trees are destroyed.

PAPYRUS 6:1   No fruit nor herbs are found ...

EXODUS 10:15   ... there remained not any green thing in the trees, or in the herbs of the fields, through all the land of Egypt.

PAPYRUS 6:3   Forsooth, grain has perished on every side.

PAPYRUS 5:12   Forsooth, that has perished which yesterday was seen. The land is left over to its weariness like the cutting of flax.

Also see Papyrus 6:1 above.

The last plague was described in Exodus 12:29: "And it came to pass, that at midnight the Lord smote all the firstborn in the land of Egypt, ..." The Papyrus said in 4:3 and in 5:6, "Forsooth, the children of princes are dashed against the well." Velikovsky said that the Bible described a series that could be explained by natural events except for the last plague where a supernatural event was inserted. He said, "An earthquake that destroys only the firstborn is inconceivable, because events can never attain that degree of coincidence. No credit should be given to such a record" (Ch. 1, »"Firstborn" or "Chosen"«). He noted that there is only a very slight difference in the spelling in the original language of "my chosen" and "my firstborn". Therefore, the original probably said, "at midnight, the Lord smote all the select of Egypt" in the way one would say all the flower of Egypt or the chosen of Egypt. (Some recent investigators consider this less likely because of the distinct difference in sound between the two words in the original.) This usage would then be identical to the prophecy of Amos when he said that during the reign of Uzziah, the select and the flower of the Jewish people shall perish as perished the chosen, the strength of Egypt.

The papyrus also supports the Exodus descriptions of the population revolting, the wretched or the poor fleeing and the king perishing under unusual circumstances.

In Egypt, after the events, the papyrus said the roads were impassable, the realm was depopulated, and there were considerably fewer people. It also said the "dwellers in marshes" and "poor men" fled the land. To make the situation worse, "A foreign tribe from abroad has come to Egypt." With Velikovsky's reconstruction, the invaders would have been the Hyksos.

In the 19th century, a traveler found a black granite shrine inscribed with hieroglyphics in el-Arish, a town on the border between Egypt and Palestine. The strange text was regarded as mythological although kings, residences and geographical places are named. The text also describes an invasion of foreigners. The events relate to King Thom and his successor and the name of King Thom was written in a royal cartouche. That indicates an historical text.

The text describes darkness during upheaval in a way similar to Exodus. A great wind was mentioned in both documents. The thick darkness was the ninth plague. The last day of darkness was when the Israelites were at the Sea of Passage. Hebrew sources say that cities were devastated in the darkness and many Israelites were killed during that time.

Both stories describe a similar time of darkness, and in both stories the death of the pharaoh in whirling waters is also similar; however, Velikovsky said that similarity is not identity. To be closer to identity, he wanted some detail of both versions that could not be attributed to chance. He found that in the name of the location. The pharaoh was killed in the whirlpool at "Pi-Kharoti." Exodus 14:9 said the Egyptians camped by the sea at Pi-ha-Khiroth [Khiroth]. The shrine also says that the pharaoh who died in the whirlpool was Thom or Thoum. Pi-Thom means "the abode of Thom." Pithom was one of the two cities purportedly built by the Israelite slaves for the Pharaoh of the Oppression.

Velikovsky said that the previous information implies two questions. How widespread was the catastrophe and when did it happen. He answered the first question in *Worlds in Collision*. He addressed the second question in the remainder of *Ages in Chaos*.

Experts agree that the *Papyrus Ipuwer* is a copy from an older document. The spelling is similar to literary texts from the Middle Kingdom. There are two periods in Egyptian history that may have a civil war and an Asiatic occupation of the Delta. The first is the dark age that separates the sixth dynasty from the eleventh dynasty, which is the separation between the Old Kingdom and Middle Kingdom. The Second is the Hyksos period between the Middle and the New Kingdoms. Experts did not agree which of those two was most appropriate. One expert said that the idea that the papyrus referred to the Hyksos probably had the most historical support, but philological considerations made him want to put the date as far back as possible. The language was of an earlier period than the New Kingdom and the text referred to establishment of "Great Houses" (law courts), which became obsolete in or soon after the Middle Kingdom. One expert (before 1952) said it was better to leave the question open for the time.

Velikovsky reopened the question. He said that both authorities had good arguments for their claims and both were partially right and partially wrong. Velikovsky provided support for the claim that the papyrus "was composed immediately after the fall of the Middle Kingdom at the very beginning of the Hyksos period." (Ch. 1, »Two Questions«) He described extensive reasons for that view.

(Comment: In *Pensée, Velikovsky Reconsidered III*, 1973, p. 36, Professor Lewis M. Greenberg noted that in 1964, twelve years after Velikovsky published *Ages in Chaos*, John Van Seters analyzed the *Papyrus Ipuwer*. He arrived at a date for the papyrus identical with Velikovsky's, but without mentioning Velikovsky. Van Seters said the last word had not been said about the dating, but "To the present writer it seems that the burden of demonstration rests on those who would still maintain an early date." Greenberg then said, "Indeed, the last word has not been said, nor is it likely to be said, so long as Velikovsky's work is ignored: in 1952 he proposed an answer to the questions Van Seters had not yet even asked.")

At Speos Artemidos is an important Egyptian inscription from the time of Queen Hatshepsut. She ruled two or three generations after the expulsion of the Hyksos, identified by Velikovsky as the Amu, from Egypt. Translations of the document say that the abode of Mistress Qes (Ques) had fallen in ruin and the earth swallowed her beautiful sanctuary and that children played over her temple. Hatshepsut said she rebuilt the temple. The inscription also said, "For there had been Amu in the midst of the Delta and in Hauar (Auaris), and the foreign hordes of their number had destroyed the ancient works;" (Ch. 1, »Two Questions«) Velikovsky said the inscription was about a natural disaster. Mere destruction by the Hyksos would not have been described as the earth swallowing the temple. In addition, the Hyksos took over Egypt and completed the destruction. He concluded that the inscription was a historical account of the time of the Exodus. Hatshepsut was recounting the history of why the temple was destroyed and was discussing how she rebuilt it. The rebuilding would not have occurred during the time of the Hyksos occupation of Egypt.

(Comment: Thirty years after Velikovsky's suggestion about the Speos Artemidos inscription referring to a natural disaster, Dr. Hans Goedicke of the Department of Near Eastern Studies, Johns Hopkins University, made the same claim, without mentioning Velikovsky. Goedicke seems to have assumed the disaster occurred not long before the inscription was written. He concluded that the inscription was written to describe the Exodus that occurred near the time of Hatshepsut and was caused by the eruption of Thera.)

A question that would assist in determining if the claim were correct is who were the Hyksos? Velikovsky takes clues from various ancient authors. One clue was that the Israelites met the Amalekites soon after observing the destructive flood at the Sea of Passage, or Pi-ha-Khiroth. Other sources say that Syria and Egypt came simultaneously under domination by the Amalekites, who escaped from Arabia when it had the same series of events described as the plagues. Manetho said the Hyksos were extremely cruel. The Hebrew narrative is similar about the Amalekites.

Another clue comes from Psalms 78:49. That verse says that after the Hebrews left Egypt, the Lord created havoc in Egypt "by sending evil angels among them." There was no plague called the evil angles and no expression like evil angels anywhere else in the Scriptures. Velikovsky noted that sending of evil angels is "mishlakhat malakhei-roim", whereas invasion of king-sheperds is "mishlakhat malkhei-roim". The difference is one silent letter, aleph, in the sending of evil angels. He said that the first is not grammatically correct and the proper reading would be "invasion of king-sheperds". The first six king-shepherds are considered the first Hyksos Dynasty of pharaohs.

Another clue is found in Numbers 24:7 where it says "… and his king shall be higher than Agog, and his kingdom shall be exalted." Some sources say that "Agog" was the title of the Amalekite kings, and it represents here the kingdom of the Gentiles. Some sources translate Agag as Agog. To indicate the level of power, a ruler was compared to the power of the Amalatike king Agog.

Manetho indicated that the Hyksos period lasted five hundred and eleven years. In 1952, accepted Egyptian chronology presented the period as about one hundred years. Flinders Petrie and others considered that span of time too small for the enormous cultural changes during the time of the Hyksos. These schemes were called the long and short chronologies. Both had a common date of 1580 BCE for the beginning of the New Kingdom. This was supposedly because of astronomical dating based on the star Sirius or Sothis. (Comment: This issue was addressed in the supplement »Astronomy and Chronology« in *Peoples of the Sea*.)

Velikovsky listed a number of similarities between the Hyksos and the Amalekites. He concluded, "This identity, established on a large number of correlations and parallels, is the answer to the two-thousand-two-hundred-year-old riddle: Who were the Hyksos?" (Ch. 2, »Hyksos and Amalekite Parallels«).

After identifying the Hyksos as equivalent to the Amalekites, Velikovsky analyzed the events following the expulsion of the Hyksos-Amalekites in Egypt and Palestine to see how they coincide. The correlation was additional support for the equivalence of the Hyksos and Amalekites.

One of the correlations he analyzed was Solomon and Queen Hatshepsut of Egypt. In the *Jewish Antiquities* of Josephus, the story of the Queen of Sheba is started with, "Now the woman who at that time ruled as queen of Egypt and Ethiopia was thoroughly trained in wisdom and remarkable in other ways, and, when she heard of Solomon's virtue and understanding, was led to him by a strong desire to see him which arose from the things told daily about his country." The problem was that the accepted chronology has Solomon and Queen Hatshepsut about six-hundred years apart.

Velikovsky listed many details that matched from the story of the Queen of Sheba and from Egyptian inscriptions about a trip of Queen Hapshepsut to the land of Punt. The list of gifts given and received by Solomon to the Queen of Sheba was nearly identical to the gifts given and received by Queen Hapshepsut. Descriptions of some of the gifts were the same in the Hebrew and Egyptian sources. The words, "never was seen the like since the world was," or "unto this day" are alike in both sources. In addition, "The rare trees, the myrrh for incense, the ivory, the apes, the silver and gold and precious stones were enumerated in both records, the hieroglyphic and the scriptural." (Ch. 3, »"The Desire of the Queen of Sheba"«) Velikovsky included reproductions of the Punt Reliefs.

Thutmose III ruled Egypt after Hapshepsut. He became the greatest of all the conquerors of the New Kingdom. Records of his successes were cut in hieroglyphs on the walls of the great Amon temple in Karnak. One of those records was about conquering Palestine.

Rehoboam ruled in Israel after Solomon. Second Chronicles 12:2 said, "And it came to pass that in the fifth year of King Rehoboam, Shishak king of Egypt came up against Jerusalem, because they had transgressed against the LORD."

The Scriptures and the Karnak inscriptions provide nearly identical details about the events. Velikovsky provided many detailed comparisons of the two stories. Many of the city names can be identified. Items that were in the temple of Solomon were depicted in the bas-reliefs of Karnak. Again, there is a six-hundred year difference between the accepted chronology and the chronology required for Shishak to be Thutmose III. This leaves Thutmose conquering a land that, according to conventional chronology, would have been inhabited by unskilled nomads. He would have had to have invaded Canaan and conquered cities built much later and taken booty which must have been copied in form and number six-hundred years later by Solomon.

Ras Shamra, in northern Syria, is an important archeological site. Velikovsky used the information to help determine the chronology of Minoan and Mycenaean cultures. He made three major conclusions. The first was that the time tables of Crete (Minoan ages) and of early Greece (Mycenaean ages) are displaced by the same time span that the Egyptian dates are out of step with the time of Solomon. The second was writings that ascribed the origin of many biblical texts to late centuries and to foreign influences are as erroneous as the reversal of it that assumes a borrowing of many biblical texts and institutions from the Canaanites of the fourteenth century. Third, he concluded that the Hurrian language is the same as Carian and a Hurrian nation did not exist.

(Comment: Part of this work led Velikovsky to suggest at a lecture at the Princeton Graduate College in October of 1953 that the Minoan B script writings unearthed on Crete are Greek. In November of 1953, Michael Ventris announced results of his initial decipherment of Linear B. Fifty of the world's most noted Hellenist scholars previously canvassed by Ventris replied unanimously that they did not expect the script to be Greek. Ventris also did not expect it to be Greek, but he proved it to be an ancient form of the Greek language.)

The remainder of *Ages in Chaos* is devoted to analysis of the el-Amarna Letters. These were mainly letters exchanged between two successive kings of Egypt and their correspondents, which were the free kings of territories in the Middle East and Cyprus and various vassal kings and princes or officers in Syria and Palestine. The letters were found at a site by the bank of the Nile where the Akhet-Aton stood. The letters contained the names of princes and governors in Syria and Palestine and the names of cities and walled places. Before 1952, no other investigator had identified any of the personnel names and only a few geographical names. Using his correlation of Hebrew with the Egyptian history, Velikovsky was able to identify some people and geographical names.

After presenting the details, Velikovsky noted that in order to keep the accepted chronology and, therefore, claim that the el-Amarna letters were written to and from archaic Canaanite princes, you must also claim that events occurred in Canaan and the events recurred about five-hundred years later in the time of Jehoshaphat and Ahab. Velikovsky said, "This makes it necessary to hold that there already was a city of Sumur, of which not a relic remained; that this city, with a royal palace and fortified walls, was repeatedly besieged by a king of Damascus, who had a prolonged dispute and recurrent wars with the king of Sumur over a number of cities, in a conflict that endured for a number of decades; that on one occasion the king of Sumur captured the king of Damascus but released him; that on the occasion of a siege of Sumur by the king of Damascus the guard attached to the governors succeeded in driving away the Syrian host from the walls of Sumur; that on the occasion of another siege of Sumur the Syrian host, hearing rumors of the arrival of the Egyptian archers, left their camp and fled – every detail an exact image of what happened again half a millennium later at the walls of Samaria." (Ch. 8, »Conclusions«)

Those are just of a few of the events that must have duplicates about five-hundred years apart. Many others were given. A few are listed here. Rimuta being the place in dispute between the king of Damascus and the king of Sumur, and Ramoth was the city in the duplicate. The king of Sumur had a second residence where the deity Baalith was worshipped and that was the name of the deity introduced by Jezebel later. The king of Damascus organized a number of ambushes against the king of Sumur, and the king of

Sumur managed to escape death each time. This was like the king of Samaria of the duplicate period. In addition, the events happened at a time when the land of Sumur had a drought and severe famine followed and the drought lasted several years and caused starvation of the people and epidemics among the animals just as in the duplicate.

Velikovsky summarizes coincidences for another two pages. If someone is willing to accept all those as coincidences then the old difficulties with the accepted chronology reappear. For example, "If the Habiru were Israelites, why then in the Book of Joshua, which records the conquest of Canaan, and in the letters of el-Amarna are no common name and no common event preserved?" (Ch. 8, »Conclusions«)

The following information is in Velikovsky's *Peoples of the Sea*, but concerns the time period discussed in *Ages in Chaos*, and provides independent support for one of the components of his revised chronology.

For twelve years, Velikovsky tried to have someone carbon date items from the New Kingdom and especially from the Eighteenth Dynasty. He finally succeeded in having the Cairo Museum provide three small pieces of wood for testing from the time of King Tutankhamen. The work was done at the University of Pennsylvania. Accepted chronology says that King Tutankhamen died in ca. 1325 BCE. Velikovsky's chronology puts the death about 835 BCE. The carbon dating indicated 1030 BCE and a test by Libby indicated 1120 BCE.

Carbon dating of trees provided a date of when the tree was cut, and not when it was used for whatever purpose. Also, for long lived trees, the date is a function of how far the sample was from the center of the tree. Long-lived cedar of Lebanon could make the results give dates older than the real date, but would not make the dates younger than the real date. Testing short-lived material should provide more accurate dates. In 1971, the British museum tested the reed of a mat and kernels of a palm from the King Tutankhamen finds. These samples provided 846 BCE and 899 BCE respectively.

By the time Velikovsky published *Peoples of the Sea*, if additional tests were conducted, the results were not published.

(Comment: If you are certain of the answer, you do not want to conduct a test in case it would give the wrong answer. If you do conduct the test and the results are not what you want, you do not publish the result.)

I have listed only a few examples from Velikovsky's extensive discussion about this period. Since I could only mention a small amount of the material, nothing I presented was meant as proof. I tried to present an idea of what he concluded and a few of the reasons why. I may have misinterpreted some of the areas, and any of the mistakes are mine. To have a more detailed understanding of the reconstruction, you should read the original books.

Overview of *Ramses II and His Time*

The Babylonians and the Medes defeated the Assyrians in the early six-hundreds BCE. Egypt was on the side of the Assyrians. The last chapters of the Books of Kings, Chronicles and Jeremiah provide many details about the events. The scriptures say the Egyptian pharaoh was Necho. They also say the ruler of Babylon was Nebuchadnezzar and mention him many times. Babylon was a powerful nation. Several Greek authors also wrote about Nebuchadnezzar. He erected grandiose buildings and archaeologists have read his prayers and building inscriptions.

The pharaoh of Egypt apparently was also powerful. He stalled the advances of Babylonians in Syria and Palestine for about twenty years. Books about Egyptian history describe Necho (II)'s wars against Nebuchadnezzar, but the information is based on the extensive data from the scriptures. Egyptian sources are almost non-existent. In one that does exist, the pharaoh is called Nekau-Wehemibre and says almost nothing about him. No records have been found about the long war with Nebuchadnezzar, the civic activities, laws created by him, temples built by him, no scrolls about him and no mummy or coffin. Velikovsky asked, "How could he have succeeded in making the Palestinian kings, Jehoahaz, Jehoiakim, and Zedekiah, believe he would be able to free Palestine from the yoke of the mightiest monarch Babylon had ever known"? (Ch. 1, »Who was Pharaoh Necho, the Adversary of Nebuchadnezzar?«) After extensive analysis, Velikovsky proposed that the monuments of Ramses II describe the same events that Jeremiah and Herodotus record about Pharaoh Necho (II).

Herodotus called the biblical Pharaoh Necho by the name of Necos. Herodotus said, among other things, that Necos defeated the Syrians at Magdalos and that he was the first to attempt the construction of the canal to the Red Sea. The construction attempt was massive and one hundred and twenty thousand workers were killed during the time. Other records indicated that Ramses II built a canal connecting the Mediterranean with the Red Sea several hundred years earlier. Historians concluded that Herodotus must have been wrong.

The activities of Ramses II are well documented. The events of the biblical pharaoh of Egypt Necho are well documented. Velikovsky compared the two sets of documents. He presented a number of similar events that are found in the documents.

Tell Nebi-Mend is near the town of Riblah and many assume it is the location of the ancient fortress called Riblah. This was the military headquarters for a time of Pharaoh Necho and then for Nebuchadnezzar. Riblah is where Necho put the king of Jerusalem, Jehoahaz, in bands and where Nebuchadnezzar blinded King Zedekiah. A fragment of a stele found at the Tell of Seti the Great, father of Ramses II, indicates that Seti built the fortress of Riblah. Accepted chronology has the fortress being built by Seti then having seven-hundred years lapse before being used by Necho and Nebuchadnezzar.

Egyptian texts describe the army as being composed of four divisions. One record mentions mercenaries of the pharaoh called the Sardan. In the scriptures, Jeremiah named four contingents of the Egyptian army. One of the four was Lydians. Sardis was the capital of Lydia. Sardan means the men of Sardis. The Lydians of Jeremiah were the Sardan of Ramses II. Ramses II supposedly ruled about seven-hundred years before the time of Jeremiah.

Velikovsky provided a comparison of descriptions of the Battle of Kadesh-Carchemish from the Hebrew sources and from the Egyptian sources. Some of the comparisons follow (Ch. 1, »The Battle of Kadesh-Carchemish«). "H" is for Hebrew and "R" for Ramses.

H: The time was four years after the first invasion of Palestine by Pharaoh Necho.
R: The time was four years after the first invasion of Palestine by Pharaoh Ramses II.

H: The place was by the river Euphrates in Carchemish.
R: The place was in the land of Khatti, Nahrin, Carchemish, Kedy, the land of Kadesh. (Kadesh had the sign of a city and not a country.)

H: The topography was described as near a fortress, surrounded on all sides by water. The fortress projected into a large stream and nearby was a sacred lake.
R: The topography was described as near a fortress, surrounded on all sides by water. The fortress projected into a large stream and nearby was a sacred lake.

H: The fortress had a double wall and moats.
R: The fortress had a double wall and moats.

H: The location of Carchemish was north of Bab.
R: The field of battle was north of Baw. (Baw is Bab of today.)

H: About the Egyptians fleeing Jeremiah said, "their mighty ones are beaten down, and are fled apace, and look not back."
R: About the Egyptians fleeing, the Annals of Ramses II said, "My numerous infantry having abandoned me, not one looking at me or my chariotry."

Some now believe that the battle at Carchemish was near Tell Nebi-Mend mentioned earlier. That site does not fit the descriptions from the Hebrew or Egyptian sources. The details of both the Hebrew and Egyptian sources describe a location by the Euphrates called the City of Jerablus (Carchemish).

The descriptions of the battle are similar in both records. The positions of the armies are the same. The army of the enemy of the Egyptians was well concealed behind the city wall. As soon as the Egyptians started to make camp northwest of the city, the enemy appeared. Part of the Egyptian army was cut off and fled north. The pharaoh and another division were attacked by another force emerging from the south of the city. Soldiers abandon the pharaoh and he heroically defended himself. He was helped by "naarim", a Hebrew word meaning boys or young men. In the time of Ahab, naarim were sometimes referred to as choice warriors. In the time of Ramses II, scribes mention naarim while discussing Palestine.

The Ramses II inscriptions usually call Carchemish Kadesh, but he knew both names. Carchemish means the city of Chemosh. Chemosh was a god and was worshipped in a large area a couple of centuries earlier. It was a holy city and the site of a large temple. The foundations of the temple have been partly excavated. Archaeologists differ on whether the site is from the seventh century or thirteenth century BCE.

The last strongholds to oppose the Chaldean army around 586 BCE were Jerusalem, Lachish and Azekah. Jeremiah 34:7 said, "... for these defended cities remained of the cities of Judah." Archaeologists found Lachish during excavation at Tell ed Duweir in southern Palestine. They also found potsherds with Hebrew lettering. The text indicated the material was from the time of the siege of Jerusalem by Nebuchadnezzar but before he destroyed Jerusalem.

Various other items were found at the same level. Some of those items were from the time of Ramses II. Scarabs and seal impressions of the Egyptian pharaohs of the Eighteenth and Nineteenth Dynasties are often found in Palestine in the same layer that is dated to the Israelite period. Explanations for this unexpected circumstance under conventional chronology are needed. Common ones are that the items are antiques used by the Israelites five or six centuries after the items were made. The other is that the items were reproductions.

(Comment: These explanations may be easy to accept if your professor says believe it or you do not receive your degree or you do not receive your grant.)

Those excuses were not used in this case. That is because the items were found in bulk and a plaque of Amenhotep III was placed under the foundation when the temple was rebuilt. Only a genuine and contemporary plaque would have been used. Archaeologists then dated the destruction of the temple about 1262 BCE. To explain the objects of the Jewish kingdom, archaeologists decided that Hebrews *dug down* into the level of Ramses II and deposited the objects.

Elsewhere in the ruins of the citadel of Lachish finds were dated to the time of Nebuchadnezzar because of Hebrew letters to the defenders of the city. At the same location, a vase with hieratic writing of the Nineteenth Dynasty and seals of Ramses II were found. To explain the objects of Ramses II at the level of Nebuchadnezzar, archaeologists decided that Hebrews *dug up* items from the level of Ramses II. If antiques and counterfeit explanations do not work, digging a hole and throwing something in or digging a hole and pulling something out are invoked as necessary.

There are many corresponding points of comparison between the military chronicles of Ramses II and with biblical records. A few are listed here:

√ At the start of the war, the pharaoh started a campaign across Palestine into northern Syria.

- √ The pharaoh met opposition in Palestine and had to fight his way through. His archers shot the opposing king.
- √ The pharaoh established a camp and an outpost at Riblah in the land of Hamath.
- √ The pharaoh took captives to Egypt from the royal house in Palestine.
- √ The pharaoh imposed a tribute on the land.
- √ Later, the pharaoh had another campaign and went to the region of Kadesh-Carchemish.
- √ The pharaoh took four divisions.
- √ Mercenaries from Sardis were in the pharaoh's army.
- √ Cities of northern Syria were allied with the pharaoh's opponent.
- √ The Egyptians were taken by surprise.
- √ The Egyptians were driven northward toward the river and not the direction of Egypt.
- √ The pharaoh hurriedly retreated to Egypt by a roundabout way with the remnants of his troops.
- √ As a direct result of this campaign, Palestine was conquered by the Chaldean-Akkadian forces (Hatti).
- √ The pharaoh started a new offensive to re-conquer Palestine.
- √ The land of the Philistines was the immediate objective.
- √ The pharaoh besieged, stormed and captured that area.
- √ For a while, Palestine was controlled by Egypt.
- √ The Egyptians left under the pressure of the Chaldeans-Akkadians.

This is an impressive correspondence for events that occurred over seven hundred years apart. The chronologies of the events have a similar correspondence.

"If we follow conventional history there is no account of Ramses' wars in the Scriptures and the wars of Nebuchadnezzar against Egypt are likewise not accounted for in extant records of the country on the Nile. But the wars of Ramses II correspond precisely with the biblical account of Pharaoh Necho." (Ch. 2, »Events of the War in the Scriptures and in the Inscriptions of Ramses II Compared«)

Byblos was an ancient royal city on the Phoenician coast. Byblos is a contender for being the oldest continuously inhabited city. Archaeologists found there King Ahiram's tomb. Hiram or Ahiram was a name of more than one Phoenician king.

The southern wall of the shaft to the tomb contained a short Hebrew inscription with the typical warning of do not go beyond this point. The warning was repeated and expanded on the lid of the sarcophagus. The tomb contained items from the time of Ramses II of the thirteenth century BCE. The tomb also contained Cyprian pottery dated to the seventh century BCE. A proposed solution to this problem was to have tomb raiders enter the tomb that was built at the time of Ramses II and deposit pottery from about six centuries later.

Assuming that there is a way you can convince yourself that the tomb raiders did make deposits, there is still the problem of the inscription in Hebrew. If people made the tomb in the time of Ramses II, the inscriptions were made at that time. Hebrew characters at the time of Ramses II were completely unexpected. The characters were very similar to those inscribed by other Phoenician kings on statues of their patrons, the pharaohs of the Libyan Dynasty, presumably of the tenth to the ninth centuries BCE. Investigators thought that the inscriptions were not written at the time of the creation of the statues, but later. The conventional chronology makes Ahiram a contemporary of Ramses II. That means the Hebrew script remained unchanged for over several centuries.

Art historians note that some of the artistic representations in the tomb match those found described in the scriptures of later time. One example was the four mourning women at each end of the sarcophagus. Two women were slapping their hips and the other two held their heads in their hands. One art historian noted examples of Old Testament hand slapping as a sign of grief, particularly in Jeremiah and Ezekiel.

The confusion caused by dating items from the same tomb at different times also caused confusion about who developed what alphabet when. Epigraphists had to try to explain the evolution of the Hebrew script, beginning around 1300 BCE, through about 900 BCE, then about 700 BCE, and finally to about 586 BCE. The major problem for the epigraphists was that the conventional chronology makes the starting point and the ending point contemporaneous, but they did not know that.

This same chronology problem is what created the five hundred year gap in Greek epigraphy. The Minoan ages are correlated with Egyptian chronology, whereas Greek ages are determined by archaeological information found in Greece. Velikovsky said there is no gap. The correlation with incorrect Egyptian chronology created a non-existent gap.

Cadmus came from Phoenicia but built in Greece. He is credited with the introduction of Ionian (Greek) letters from Phoenicia. Under the historical reconstruction, Velikovsky suggested that Nikmed, or Nikdem of Shalmaneser III's war records, was the legendary Cadmus. Velikovsky also suggested because of Cadmus' background, he may have experimented with Linear B script by making an alphabetic writing of it before he found the best solution in using Hebrew letters for writing Greek. At an early palace in the Greek Thebes, Cadmeion, archaeologists discovered roll cylinders with cuneiform signs. Translating some was difficult. Velikovsky suggested, "An effort to read them on the assumption that they are in cuneiform alphabetic Greek may, perchance, prove successful." (Ch. 3, »The Inscriptions of Ahiram and the Origin of the Alphabet«)

Archaeologists excavated various sites around Byblos. They found a number of objects with the name of Ramses II, who supposedly lived around the middle of the 1200s BCE. They also found items mentioned earlier from around 500 to 600 BCE. The results of the excavations at Byblos presented an interesting fact that caused numerous discussions. Velikovsky quoted Nina Jidejian as saying, "In the excavated area at Byblos there is a complete absence of stratified levels of the Iron Age, that is for the period 1200-600 B.C." (Ch. 3, »"A Curious Fact"«) The reason is obvious if the chronology is wrong.

The great pharaoh Necho started work on a canal which would ultimately connect the Mediterranean Sea and the Indian Ocean. He sent an expedition around Africa. He waged great wars and impressed Greek authors and Jewish authors. He also did not leave any Egyptian records of his achievements. Ramses II did leave records of the same events, but nobody else seemed to notice. Nebuchadnezzar was a powerful king in the time of Necho. Nebuchadnezzar also waged great wars and impressed Greek authors and Jewish authors. The historical records of Nebuchadnezzar seem not to exist. Velikovsky said, "It seems odd that a great and long war between Egypt and Babylonia, recorded in such detail in the Scriptures, should have been non-existent in the records of the main participants." (Ch. 3, »A Recapitulation«). He also said that since we now know the real time of Ramses II, we can determine what misplaced records apply to Nebuchadnezzar. Chapters IV through VII provide details of that investigation.

Peculiar pictograph inscriptions have been found near Ivritz, on the plateau of Asia Minor. Similar pictorial signs are carved on stone that had been re-used in a building at the bazaar of Hamath in northern Syria. The signs were also on slabs in the area of Jerablus-Carchemish on the bank of the Euphrates. The signs were also found at the site of ancient Babylon and in other places. The signs are completely different from Egyptian hieroglyphics. The origin of these signs was not known.

Rivals of Ramses II were mentioned in the bas-reliefs of the battle of Kadesh. The rivals are called Kheta. They are also mentioned in the poem celebrating the battle and in the Egyptian text of the peace treaty between Egypt and Kheta. In 1870, because of a phonetic similarity, historians concluded that the Kheta must be the Hittites who were occasionally mentioned in the Scriptures. Since almost nothing was known about the Hittite history, some called it the "forgotten empire." Not everyone agreed. Some considered it strange that the ancient world of the empires of Egypt and Assyro-Babylonia should be increased by a newly discovered empire of the Hittites.

The ancient ruins of Boghazkoi are in Turkey about one hundred and forty kilometers east of Ankara. Numerous tablets were found there. Some were inscribed in the Babylonian (Akkadian) language. Others had cuneiform signs that were not recognized.

One of the tablets in Babylonian was a copy or draft of the treaty between Ramses II and the king of Hatti, Hattusilis, in the cuneiform and Khetasar in the Egyptian. The treaty had an Egyptian version found in Egypt. In the hieroglyphic text, Khetasar is called "the great chief of Kheta," and in the cuneiform text, "the great king of Hatti." This was thought to confirm the Hittite empire, but there were problems. For example, the tablets were found in a layer much more recent than the assumed age of the documents.

The tablets seemed to have at least three main languages and several secondary languages. There was Babylonian, a dialect called Hittite by the investigators, a tongue called "the language of Khatti," and four or five other languages. Velikovsky said that based on the placing of the tablets in the seventh and sixth centuries before the present era, it may be possible to identify some of the languages as Chaldean, Phrygian, Lydian, Median and Etruscan. Modern scholars determined that Lydian "seems to be Hittite." This would be reasonable under the revised chronology. Other investigators found that the "Hittites" had scholarly works, hymns, writings based on historical traditions, vocabularies, and other literary works in common with the Babylonians.

The Assyrian Empire is supposed to have started ascending after the fall of the Hittite Empire. Since the Hittites appeared more advanced than the Assyrians, historians thought the Assyrians regressed culturally compared to the Hittites. Current chronology has the Assyrian Empire starting not long before about 1100 BCE, and the Hittite Empire existing before that. Under the revised chronology, no regression is needed to explain the reduction in culture of the Assyrians. They came before the "Hittites".

Two long inscriptions found at Boghazkoi were versions of Mursilis' war annals. For a number of years, Mursilis fought against a coalition of the king of Assyria and the King of Egypt. The war continued without a conclusion and in the ninth year the war was waged in Harran. There Mursilis fought the king of Assyria, Assuruballit. Mursilis was the father of Hattusilis. In the revised chronology, this would make Mursilis equivalent to the Chaldean Nabopolassar who was the father of Nebuchadnezzar. To determine if the correlation was reasonable, Velikovsky compared facts found in the annals of Mursilis with the facts known about Nabopolassar, king of Akkad (Babylonia) and Chaldea.

The following is a sample of the correlations:

√ Mursilis marched along the Euphrates and battled the Assyrian troops supported by the Egyptian troops.
√ Nabopolassar marched along the Euphrates and battled the Assyrian troops supported by the Egyptian troops.
√ Mursilis' military operations against Assuruballit were in Harran.
√ Nabopolassar's military operations against Assuruballit were in Harran.
√ Mursilis' last fragment of the war annals was in his twenty-second year.
√ Nabopolassar died in the twenty-second year of his reign.

Under conventional chronology, the events are about seven hundred years apart. No other period in history had Assyria and Egypt as allies in a war.

To help in understanding how confusion is created, Velikovsky described naming customs of the times under discussion. For various reasons, it was the custom on some of the areas for a survivor to take the name of a deceased person. Like the Egyptian pharaohs and Jewish kings, the princes and kings of Assyria and Babylonia had more than one name. Egypt had a statute that the king should have five royal names. Not all the names were permanent and were occasionally replaced. Ramses III had more than a dozen names. Sometimes a person was referred to by different names in the same story. A king would also be called by different names in different provinces. It was also common to change a name of a person so that it sounded more agreeable to the ears of foreign people. In addition, different gods had to be honored by people taking associated names. Velikovsky said, "Besides all this, cuneiform can be read both ideographically and syllabically, and thus 'Negril' (Nergal) could become 'Muwatallis'." (Ch. 4, »Names and Surnames«) This makes it understandable how the Greeks, Hebrews, Assyrians, and Egyptians could tell the same story and it could appear to be about different people. Confuse the chronology and you are forced to make the story about different people at different times.

A detail demonstrates how archaeology and texts do not match but the physical evidence may be ignored. Nebuchadnezzar spent much time and money on the temple of Esagila in Babylon. The bricks in the foundation have his name. The walls are still impressive. Nebuchadnezzar rebuilt the temple from the foundation to the roof and covered it with gold. According to Nergilissar's inscriptions, at a time considered to be two years after Nebuchadnezzar's death, Nergilissar made repairs to the ruined temple structure and covered the gates with silver. Velikovsky notes that the temple structure would not be ruined in two years with no mention of a cause. That and many other details from Greek and Babylonian sources led Velikovsky to conclude that there was possibly an earlier Nergilissar I who preceded Nebuchadnezzar. The succession would then be Nabopolassar, Nergilissar I, Nebuchadnezzar. Evil-Marduk was next and then another King Nergilissar came after that.

Nearly every "Hittite" find excavated in Anatolia and northern Syria could be interpreted as belonging to two different ages. Phrygian mound tombs with antiquities were found in Gordion and determined to be seventh and sixth centuries BCE. Another researcher said that the items were more than six hundred years earlier. This was because excavators at Alisar, about eighty kilometers away, assigned a stratum with similar finds to the fourteenth and thirteenth centuries BCE. The stratum had items with "Hittite" pictographs. Someone else complained that that opinion was wrong because some items in the stratum were from a much more recent time. Researchers were so confused some said the statements about depth in determining age were worthless. This allowed them to find Greek objects from around 600 BCE in the stratum dated around 1200 BCE and not consider that strange.

To complicate matters more, the Hittite hieroglyphs were found only in the one stratum at Alisar. This led one researcher to claim that the hieroglyphs may not be Hittite. The level was renamed the "first post Hittite level." Later, a radiocarbon test was made on a piece of wood found in what originally was labeled Old Bronze Age. The result was that the layer was about seven hundred years more recent than thought based on conventional chronology. Velikovsky quoted a leading German Hittite authority as saying that evidence indicates that Hittite history has no chronology of its own, and "we must build the Hittite chronology upon the Egyptian chronology." (Ch. 6, »"In the Deepest Darkness"«) (Quote given from Götze, *Mitteilungen*)

No Phrygian artifacts have been found that date before the first half of the eighth century BCE. The end of the Phrygian kingdom is known to be around 687 BCE. Under conventional chronology, the Hittite level should be found at Gordion beneath the Phrygian level. Velikovsky's revised chronology would expect some Hittite artifacts above the Phrygian layer. This is what was found. To make the data fit the desired explanation, one researcher said that the abundant presence of Hittite relics occurred because someone brought in the Hittite layer from somewhere else and deposited it on the Phrygian layer. The Hittite layer averaged about four meters thick.

Large areas of Asia Minor have no artifacts from Anatolia from about 1200 BCE to about 750 BCE. There is nothing from Phrygian or any other people. This is certainly not a surprise with respect to the revised chronology.

Beneath the floor of a room of the Northwest Fort of Carchemish, excavators found a tomb with golden objects. The tomb is called the Gold Tomb of Carchemish. Among other things, thirty-nine small figurines were found that were reproductions of the great rock-cut reliefs of Yazilikaya. The problem is that carvings are of the thirteenth century BCE and the grave is from the last years of the seventh century BCE. Two theories were presented. One was the heirloom theory that said the items were made in the thirteenth century but passed on from generation to generation. The other was that the items were made in very late Hittite time and were only a few hundred years older than the tomb. Both theories had problems. One historian said that a third possibility would be to doubt the age of the tomb itself. Velikovsky proposed another solution. He said, "The reliefs of Yazilikaya are not of the thirteenth century but younger by six to seven centuries" (Ch. 6, »The "Gold Tomb" of Carchemish«). That explains how carvings matching the Yazilikaya reliefs are found in a seventh century tomb.

Malatya is at the center of the mountainous region of eastern Anatolia. Art historians and archaeologists have great interest in a Lion Gate there. Art historians say that the art is obviously closely related to Hittite art. The conclusion was that Malatya was of the Hittite Empire time. One of the arguments was that the gate had an obvious connection to the art of Yazilikaya. This meant Malatya was associated with the twelfth century. Archaeological evidence indicated the gate was built in the mid-eighth century. Similar to the above paragraph, Velikovsky's solution explains how art matching the Yazilikaya reliefs are found on a seventh century Lion Gate.

Nebuchadnezzar mentions a colony of Greek soldiers while describing his visit to Egypt. The Hebrew name for the city is Tahpanheth. It was a frontier town east of the Delta. The Greek name was Daphnae. Greek soldiers were stationed there during the seventh and sixth centuries BCE. The location was selected for protection of the Palestinian border of Egypt. The city was supposed to have been built around 664 BCE and lasted until about 565 BCE.

Excavators there discovered amounts of Greek armor, tools and wares. Also, the excavators found foundations of a temple built by Ramses II. In addition, the ruins contained part of a statue. The statue part contained cartouches of Ramses II. The Ramses II items were unexpected.

Reddish kiln-baked bricks were found there and in a neighboring village of Nebesheh. In the temple at Nebesheh, a statue with the cartouches of Ramses II was found. Also, Jeremiah 43:9 says, "Take great stones in thine hand, and hide them in the clay in the brick kiln, which is at the entry of Pharaoh's house in Tahpanheth, in the sight of the men of Judah." Other data make it appear the baked bricks were not discovered in Egypt earlier than the Ramesside period or of most ages following that time. Nebuchadnezzar rebuilt his father's palace by replacing the walls of mud brick by baked brick. Nebuchadnezzar refers to the baked bricks often in his building inscriptions. Using conventional chronology, it appears that the kiln stood for seven centuries from sometime after Ramses II down to Jeremiah without being used in the interval and that the bricks made in the kiln in the days of Jeremiah all vanished.

The rest of the last chapter concerns remnants of the people who were crushed between Nebuchadnezzar and the Egyptian pharaoh when they were at war and then when they became friends.

Following that is an epilogue with eight questions and answers. For example, question 6 states: "Stratigraphy dominates all judgments of professional archaeologists. Literary monuments are considered of definitely secondary value and when found in wrong stratigraphical positions, are considered to be intrusions. Pottery, however, especially Mycenaean and post-Mycenaean (Geometric in various stages, and Orientalizing), defines by its presence the chronological placement of the strata. Scarabs, often carrying royal Egyptian names, are second only to pottery (usually sherds) as arbiters of age. What is the verdict coming from pottery and scarabs in the court where conventional chronology and the revised scheme of it stand before the bar"? (Epilogue)

Velikovsky explains the answer in nearly ten pages of information in the section titled "Scarabs and Stratigraphy." Part of that text describes finds such as one in a tomb where a large collection of scarabs all point to a particular period but for some reason the tomb is given another age six hundred years earlier. The accepted conclusion is that grave robbers stole the collections and buried it in the tomb at a later period. This attempt to support conventional chronology and other data Velikovsky supplies demonstrates that the pottery and scarab data may reasonably fit the revised chronology, but that does not mean that a biased group will render the correct verdict.

Overview of *Peoples of the Sea*

In *Ages in Chaos – From the Exodus to King Akhnaton*, the first volume in Velikovsky's discussion about the chronology of Egyptian history, he presented evidence that placed the Exodus at the end of the Middle Kingdom in Egypt. He moved the time of the Middle Kingdom closer to the present by more than five centuries. His total reconstruction covered a span of twelve centuries, from the end of the Middle Kingdom at least to the arrival of Alexander the Great of Macedon. *Peoples of the Sea* deals with nearly two centuries of Persian domination of Egypt and continues through the conquest of Egypt by Alexander the Great and the time of the early Ptolemies. He considered the reconstruction described in this book as the largest change from the accepted chronology.

As with *From the Exodus to King Akhnaton*, although Velikovsky took personal responsibility for his books, again he did not work in a vacuum. Velikovsky always consulted experts in various fields before publishing a work. *Peoples of the Sea* was no exception. For example, Egyptologist Dr. Walter Federn supplied much bibliographical information. Dr. N. B. Millet, curator of the Egyptian Department of the Royal Ontario Museum in Toronto read proofs of the book and provided useful suggestions. Dr. Martin Dickson, Professor of Oriental studies at Princeton University and Professor R. J. Schock, at the time chairman of the Department of Classical Studies at Brooklyn College also read the proofs and provided encouragement for the work. Many other scholars such Professors Lynn E. Rose and Lewis M. Greenberg, provided Velikovsky with leads to important information.

(Comment: In a discussion of what is in *Worlds in Collision* and *Earth in Upheaval*, the reader should keep in mind that there are two major components to Velikovsky's works about physical events in the recent history of the Solar System. The components are (1) data demonstrating that the generally accepted (1950) theory had problems, and (2) a model using parts of that and additional data to develop a new model. The same is true for the historical works. The immediately following information, which was in the supplement to *Peoples of the Sea*, was about problems with the accepted (1950) Egyptian chronology. The first chapters of *Peoples of the Sea* contained both problems with the accepted chronology and suggestions about a model that would solve some the problems. Those chapters will be discussed after the supplement.)

A supplement to *Peoples of the Sea* is called "Astronomy and Chronology." It is a major contribution to the discussion about why Egyptian chronology is not as well established as many Egyptologist believe. Because of its importance to the overall reconstruction, I will cover it first in this overview. Professor Lewis M. Greenberg believes that *Ages in Chaos* would have had a greater impact if "Astronomy and Chronology" had been an introductory section to *Ages in Chaos*.

(Comment: Although advertised as known and fixed, Egyptian chronology has had some minor changes through the last fifty years. The dates given here are some considered correct in 1950. For example, dates here for Ramses I start at 1321 BC. Current guesses frequently cited are 1306-1304 BC, or 1292-1290 BC.)

Ancient history in the Middle East, especially the history of the second millennium before the present era, is related to the chronology of Egypt. You can determine a relative chronology for a country, but to determine real dates, you need to link somewhere in the sequence to Egypt either directly or through a third region. For example, the lawmaker king Hammurabi of the First Babylonian Dynasty was placed about 2100 BCE. Material from the time of Hammurabi and from the Egyptian Middle Kingdom was found in a deposit on Crete, so the date for Hammurabi was changed to about 1700 BCE to synchronize with the Egyptian dynasty.

Obviously, it is extremely important to have Egyptian chronology correct to ensure that the chronologies of other countries are correct. Velikovsky questioned how strong the support is for the current Egyptian chronology. He noted that the question is being asked a little late since an extremely large amount of literature has been composed based on the accepted chronology. It is apparent that anyone devoting their life to histories tied to the accepted chronology would be unwilling to consider a new chronology.

Velikovsky said, "Everyone is agreed that Egyptian chronology is so well devised, century by century, decade by decade, and often year by year, that no new evidence could break down this massive growth. What, then is the foundation of this system, which the Egyptologists have concluded is absolutely firm and from which scholars in other fields have confidently borrowed their data and standards?" (Suppl., Ch. 1, »A Mighty Tree«) He demonstrated that the foundation is not nearly as firm as many people believe.

(Comment: Compare the wording of Velikovsky's first quote with "It is thanks mainly to the Sothic theory that commentators now generally agree that Egyptian chronology is so well devised, century by century, decade by decade, and often year by year, that no new evidence could dismantle this elaborate structure." That sentence is from *Sothic Dating Examined: The Sothic Star Theory of the Egyptian Calendar* (A Critical Evaluation) October, 1995 Sydney, Australia. It was written by Damien F. Mackey (MA. B PHIL) for a Masters degree. Mackey did refer to Velikovsky's *Peoples of the Sea* in other locations in the thesis. Unlike Mackey, you could expect many of my phrases to resemble Velikovsky's because I am writing an overview of what he said.)

There is no known Egyptian system that shows continuous counting of years from a specific time. Egyptians dated according to the number of years the current ruler had been in power. For example, Hatshepsut visited the Divine Land in the ninth year of the Queen. If there were co-rulers, it is not always easy to determine how those years are counted. It is not always possible to establish succession of kings from a given dynasty. The correct sequence of dynasties is not definitely determined.

Some early documents have the genealogy of kings, but the list does not extend to the New Kingdom, which is of importance here. Some lists seem exaggerated because the Thirteenth Dynasty, last of the Middle Kingdom, listed over a hundred names. This at best makes the documents questionable.

Velikovsky noted that some of what is assumed about the sequence of events in Egyptian history came from Manetho, an Egyptian writer, historian and polemicist. He apparently was an anti-Semite because he invented the baseless identification of Moses with Typhon, the evil spirit, and the Israelites with the hated Hyksos. Manetho contradicted himself by identifying Moses with the rebellious priest Osarsiph of much later times. In addition, Manetho wanted to demonstrate that the Egyptian culture was much older than the Greeks who ruled Egypt at the time. This makes much of Manetho's sequence questionable.

There are two versions of Manetho's list of dynasties. It is also not easy to determine which of the kings known from monument inscriptions are meant by Manetho. Sequences of kings with strange names are never found on monuments. Velikovsky quoted Breasted as saying "The chronology of Manetho [is] a late, careless and uncritical compilation, which can be proven wrong from the contemporary monuments in the vast majority of cases, where such monuments have survived" (From H. R. Hall) (Suppl., Ch. 1, »"A Skeleton Clothed with Flesh"«). Where monuments existed to prove the lists were wrong, the lists were changed. Where there were no monuments, the lists remained unchanged as if from a reliable source. Ramses III had information about him on monuments, but he was not mentioned in Manetho's lists. Egyptologists assigned Ramses III to the Twentieth Dynasty possibly because Manetho did not have names for kings in that dynasty. A list by Georgius Syncellus did have names for that dynasty, and Ramses III was not on it. One of the lists had the Twentieth Dynasty being 135 years and the other 178 years. This gave some room for adding people as necessary to retain something similar to the lists.

The same person quoted above who said that Manetho's lists were "late, careless and uncritical compilation" also said "The main scheme of the history of ancient Egypt is now a certainty, not a mere hypothesis; but it is very doubtful if it would ever have become a certainty if its construction had depended entirely on the archaeologists" (Suppl., Ch. 1, »"A Skeleton Clothed with Flesh"«). Hall was referring to help by astronomers. Velikovsky demonstrated that the archaeologists may have done better without the incorrect constraints placed on Egyptian history by astronomers. Not only that, he said, "But actually it was not the archaeologists who originally filled out the scheme of Manetho with data derived from hieroglyphic texts chiseled on monuments or written on papyri. The strange fact is that long before the hieroglyphs were read for the first time the kings of Egypt were placed in the centuries in which conventional chronology still keeps them prisoner" (Suppl., Ch. 1, »"A Skeleton Clothed with Flesh"«).

Before addressing the Sothic dating issue, Velikovsky examined the history of who first placed Ramses III in the twelfth century. He noted that historians placed Ramses III in the twelfth century before Champollion deciphered the Rosetta Stone. Therefore, historians made the placement before any monumental inscriptions were read to support such a placement.

Two years before the translation of the Rosetta Stone, a book was published that said Ramses III started his reign in 1147 BCE. Another writer said the date was 1477 BCE and another said 1279 BCE. None gave any support for their date. No reference is known to exist where Herodotus or Thucydides or other classical authors mentioned Ramses III. Modern scholars gave the name Ramses III to the person associated with the impressive temple of Medinet Habu. When information from Medinet Habu was finally deciphered, it revealed that the king fought the Philistines, which fits well with the dating in the twelfth century. Philistines were a significant part of the book of Judges. Velikovsky asked if there were reasons to revise the time of Ramses III. That led into the discussion of the Sothic dating.

Historians made the dynasties listed by Manetho the framework of Egyptian history. This was even while Breasted considered the time spans "absurdly high". The problem appeared to be solved because historians used astronomical evidence to determine calculated times for various events.

No solar or lunar eclipse records were found in Egypt. It is thought that Egyptians used the Sothic period for measuring years. The Sothic period is a computation based on the rising of the star Sothis ("Spdt" in Egyptian) or Sirius or "Dog Star". The "heliacal rising" is the first day when the star rises and the Sun is far enough below the eastern horizon to make it visible in the morning twilight. This became the cornerstone for the accepted dating.

The heliacal rising of Sirius announced the flooding of the Nile.

It is known that Egyptians of the Hellenistic and Roman periods knew there were 365 ¼ days in a year, but priests did not want to use that for the calendar because it would make the feast of Isis stationary with respect to the seasons. The Roman author Censorinus said in 238 CE that Egyptians were out of sync by one year every four years. Correspondence was reestablished after 1461 years. He said that this was called a great year, the heliacal year and 'the year of the God.' Censorinus also said that Aristotle described a "supreme year," which lasted until the sun, moon and planets returned to their starting positions, and he described a "cataclysmic year," which was the period between catastrophes. According to Aristarchus of Samos, the cataclysmic year was 2484 years.

The calculation of the great year is not exact because the Julian year of 365 ¼ days is not the sidereal year. The difference between 1460 years of 365 ¼ days each and 1461 years of 365 days makes a difference of about thirty six years in a Sothic period. To explain how the Egyptians acquired their knowledge of the Sothic period by considering this difference, investigators rely on "exceedingly rare chance" of two phenomena that are detailed in the book.

Censorinus also said that a new Sothic period started one hundred years before he wrote his work. That would be about 139 CE. The previous period would be 1322 BCE. That date is the foundation of Egyptian chronology. On an old manuscript, Theon provided additional information that allowed calculating the time of Menophres at 1322 BCE.

Historians usually consider Menophres to be Ramses I, the founder of the Nineteenth Dynasty, although Borchardt considered Menophres to be King Seti the Great, son of Ramses I and father of Ramses II. Since Ramses I appears to have ruled only one year, it is easy to place him in the 1322 BCE timeframe. Establishing a chronology from the one date would have been easier but there appears to be no known documented instance of an Egyptian event being recorded by the serial year of a Sothic period. The predominant opinion is that the ancients did not consider the Sothic period as an era as a method to reckon years and that it is used by modern historians to calculate chronological dates. Only a few ancient Egyptian references have been found to help with the calculations.

The Books

One document found at the Illahun temple referred to a Sothic event that seems to have occurred in the time of either Senwosret III or Amenemhet III. This placed his seventh year at about 1877 BCE. This Twelfth Dynasty placement allowed providing dates because of sequence and span time for other rulers in the Twelfth Dynasty. Another has a reference to a Sothis rising in the time of Thutmose III from the Eighteenth Dynasty. The month and date are given, but not the year in the Thutmose III reign. Even if you assume the rising was a heliacal rising, the calculations could be off several years. Another reference has even less information. This leaves the Illahun reference to the Middle Kingdom and the Theon document as the only ones that can be counted on for constructing a chronology based on Sirius' heliacal rising, or a Sothic period of 1460 years.

With only those two that are assumed to be firm points and information about a few moon festivals, there are a number of problems in developing the chronology. For example, Josephus used information from Manetho and said that the Hyksos ruled 511 years. Because of the points thought of as firm, there is only room for about 100 years for the Hyksos. Historians thought about adding an additional Sothic period to allow 511 years for the Hyksos but decided that would cause more problems. The solution was to ignore Manetho.

Velikovsky notes that the chronology basically depends on the statements and on the correct interpretation of the statements of Censorinus and Theon. They wrote in the third century CE and had little access to original documents. In addition, Velikovsky said that some of the earlier sources used by those men were written by people not necessarily "conscious of the importance of separating fact and supposition." He gave some examples of an early Latin author who seriously discussed the phoenix visiting Egypt occasionally.

Censorinus and Theon appear to consider it appropriate to fit a 1460 year period into Egyptian chronology. The Egyptians themselves did not mention such a period.

Because a major cornerstone of the accepted chronology is the placement of Menophres, Velikovsky covered in detail the question of who was Menophres. Theon did not say that Menophres was a king. He may have held any of a number of high ranking positions. The Egyptian data do not mention an era where Menophres is found. J. B. Biot suggested that Menophres stands for Men-Nofre, the Egyptian name for Memphis (Suppl., Ch. 2, »Who was Menophres?«). Although the idea was rejected in the late nineteenth century, Velikovsky explained why it had some merit. If Menophres was a city instead of a person then there is no major point to establish the Egyptian chronology.

If Menophres was a king who ruled at the beginning of a Sothic period era, it is still difficult to identify him. In the kings lists various kings with similar names are mentioned, but none have the name Menophres. Some of those names have no data from monuments to show that they are real persons. Velikovsky details why several of the names considered would not fit. He summarized by listing assumptions used as a basis for much of Egyptian chronology.

The assumptions are: "(1) that there had been an era of Menophres; (2) that this era coincided with a Sothic period; (3) that this Sothic period began in -1321; (4) that

Menophres was a king who lived at the beginning of this epoch. In addition to all these assumptions and surmises it was maintained (5) that Menophres was Ramses I, because the beginning of the reign of Ramses II was *a priori* (and without any sufficient reason) placed at -1300." (Suppl., Ch. 2, »Who was Menophres?«)

He concluded that "The chronology of world history constructed on these hypotheses does not seem so stable and secure as was thought; it looks more like an aggregation of many unconnected things each unstable by itself, piled precariously one upon the other." (Suppl., Ch. 2, »Who was Menophres?«)

The preceding information, which was in the supplement to *Peoples of the Sea*, was about problems with the accepted (1950) Egyptian chronology. The first chapters of *Peoples of the Sea* contained both problems with the accepted chronology and suggestions about a model that would solve some of the problems. The following is an overview of the first chapters of *Peoples of the Sea*. Velikovsky said that this work can be read independently of the other volumes about the history of countries in the Middle East.

In accepted chronology, the twelfth century BCE was a time of great changes in nations around the eastern Mediterranean, also called the Ancient East. Migration uprooted populations of large areas. The Mycenaean culture ended and the Trojan War occurred. Soon after the Trojan War, armed hordes attacked Egypt, ruled by Ramses III at the time. He successfully fought the invaders in what is known as the war against the Peoples of the Sea. Ramses III called them People of the Isles. Ramses III started his reign around 1200 BCE.

Nothing is known from other sources about the Peoples of the Sea before they arrived at the borders of Egypt. Their existence is inferred because Mycenaean Greece, the Hittite empire and a number of lesser kingdoms ended about 1200 BCE. For four or five centuries after that, there is no record or relic of their existence. Many countries went into a Dark Age period. Suddenly about 700 to 750 BCE things revive. Homeric poetry appeared. Homer was familiar with the smallest detail of the life of the Mycenaean Age, over five centuries earlier. Why this is so creates extensive debates.

This time was not a Dark Age similar to that of Europe. There are relics and other documentation about that time span. In the 1200 BCE to 700 BCE span, no document survived from that time in Greece, in Crete, in the Aegean world, or in Asia Minor. Velikovsky said that this problem was created by accepted Egyptian chronology. Other problems include changing the time of the Exodus and reduce the time of the Judges from four hundred years to slightly over one hundred years.

In order to determine the real relationship between the Mycenaean, the biblical, and the Egyptian events, Velikovsky reexamined the historical material left by Ramses III. Velikovsky started with the palace at Tell el-Yahudia and then his mortuary temple at Medinet Habu.

Some of the tiles found in the palace area have Greek letters. No traces of the language have been found in Greece before about 750 BCE. It is significant the Greek letters on the Egyptian tiles do not look like early Greek of the seventh century but like the classical letters of the age of Plato. Velikovsky presented references from a number of discussions about that problem. The main suggestions about how to solve the problem under conventional chronology [at the time Velikovsky wrote the book] require great imagination and obviously are not good solutions.

There is another problem. The other side of the tiles, according to one expert, "strikingly reminds us of Persian art." The Persians arrived in the latter part of the sixth century and were expelled by Alexander around 332 BCE. How can this be? According to the Egyptologist Griffith, "The question involves a great difficulty." (Part I, Ch. 1, »Greek Letters on Tiles of Ramses III«)

Two well known archaeologists found the ancient cemetery of the site about a mile away. It contained graves built of basalt blocks and sand. "In each case, the tomb consisted of an outer case of large crude bricks; a kind of vaulted roof was made of bricks leaning against each other; inside was placed a terra-cotta coffin in the shape of a swathed mummy, made of one piece, with a large opening at the head through which the corpse was introduced, apparently not mummified. (Part I, Ch. 1, »Necropolis: Twelfth of Fourth Century?«) The head was covered with a faceplate. The features on the faceplate were in the style of late cemeteries of Erment or Alexandria.

Most of the coffins were painted in a manner such as found on mummies of Greek and Roman time. (The dead were not mummified, but mummies were painted on the coffin.) Most of the adult tombs had been pilfered, but some child's tombs were intact. One child's tomb contained a necklace of porcelain and glass beads and a ring set with a small scarab. On the breastplate was a small Cypriote vase or amphorae, an ancient Greek jar or vase with a large oval body, narrow cylindrical neck, and two handles that rise almost to the level of the mouth. A hieroglyphic on one coffin was partly readable and had a Greek ending. One tomb contained two pottery scarabs that had the name of Ramses III. One was inscribed with the name of Setnekht [father of Ramses III]. The other was of Ramses VI, the successor of Ramses III. Two graves had pottery with a few "letter like" marks, M and C. Over three hundred miles away, in the tomb of Ramses III, similar jars painted on murals had two handles.

This is a second time for the same dilemma. The data suggested a late period for the items associated with the time of Ramses III, but conventional chronology has the time of Ramses III over five hundred years earlier.

Ramses III left extensive inscriptions carved in stone and texts of papyri, the largest of which is called the *Great Papyrus Harris*. That document has the character of the king's last will and testament. It is thought to have been composed during his time by possibly Ramses IV. The document referred to the great achievements of Ramses III and his predecessor, Setnakhet.

The odd part of the text was that it mentioned Setnakhet regained power for Egypt from foreigners. Under conventional chronology, there were no foreign invaders occupying Egypt just before Setnakhet. No other documents suggest a foreign domination.

Also, the accepted chronology leaves only a few years between the end of the Nineteenth Dyansty and the beginning of the Twentieth Dynasty. The document described a long period while the Egyptians were under someone named Arsa. There is not enough time for the events described and the events do not fit the period. Arsa is not known under the accepted chronology.

Ramses III built an impressive mortuary temple to himself. The walls have engravings about his military victories. In one, the large opposition force gathered in Canaan-Palestine. Part of that force was under the leadership of a nation whose name read Pereset. Their allies were the Peoples of the Sea. Drawings of the Pereset solders depict them with crown-like helmets. Soldiers of the Peoples of the Sea have horned helmets that sometimes have a ball or disc between the horns. In excavations in various countries, the crown-like helmets are found on Persian soldiers. The tomb of Darius has bas-reliefs showing the guard of the Persian monarch. The guards have the same headgear as the Pereset. This and other information supplied by Velikovsky suggests there is a strong argument for identifying the Pereset as the Persians. Under the accepted chronology, the Persian contact with Egypt and Persian wars with Egypt are limited to between about 525 BCE and 332 BCE whereas Ramses III was closer to about 1200 BCE.

There is evidence in addition to drawings. Around 238 BCE, a group of priests met to discuss calendar reform. They made a decree known as the Canopus Decree. It was written in Greek, in demotic Egyptian (cursive), and hieroglyphics. The text refers to a time when the P-r-s-tt took sacred images from Egypt and Ptolemy III went to where the statues were and returned them to Egypt. The Greek version of the same event said the sacred images were taken by the Persians. The P-r-s-tt would be the Pereset of the inscriptions at Medinet Habu of Ramses III. (Other geographical locations in the decree also have an extra t.)

Velikovsky noted that we cannot let Ramses III fight with the Persians and keep the hinges of world history in their former place.

Because of the possible timing, Velikovsky analyzed data by Greek authors. They discussed Egyptian kings Nepherites, Acoris, Nectanebo I, Tachos, and Nectanebo II. In the conventional Egyptian chronology, those late pharaohs are in the so-called Twenty-ninth and Thirtieth Dynasties.

Data from the time of Ramses III describes his war with the Pereset and the Peoples of the Sea. Data from the Greeks describes in detail the war of Nectanebo I against the Persians and Greek mercenaries.

The Pereset first helped the pharaoh Ramses III and then became his enemy. The Peoples of the Sea helped the pharaoh and then became his enemy. The Persians helped Nectanebo I and then the pharaoh started a war with the Persians. The Greeks helped the Egyptians and then fought against Egypt.

Diodorus wrote about events of the time of Nectanebo I. Documents of Ramses III and Diodorus both describe how in order to obstruct a forced entrance to the mouths of the Nile, the pharaoh raised walls in them. Ramses III documents and Diodorus said that the enemy entered the Nile mouths and occupied a fortress. Both Ramses III and Diodorus said this was a disaster. Both described in almost identical terms the slaughter of the invaders – the Pereset and the Peoples of the Isles, or the Persians and the Greeks.

Velikovsky said the data indicate that the Greek Nectanebo I was the Egyptian Ramses III. Their personalities, their lives, their rules, and their wars are extremely similar. Ramses III even used the name Nekht-a-neb as one of his royal, so-called Horus names. The identification of Ramses III as Nectanebo I explains why the Egyptian records do not mention Nectanebo's war and the account was from the Greeks. Similarly, Velikovsky said, "Of Ramses III's war, neither Hebrew nor Greek historical data will be discovered because the record of it is the history of Nectanebo I."

There are other details that seem to indicate that Ramses III lived at a much later time than assigned by conventional chronology. In drawings of the faces on the bas-reliefs of Ramses III, an art expert said the Greek type face was very noticeable. They also did not have beards. Not until the late fifth or fourth century did Greeks begin to shave their faces. The armor of the Peoples of the Sea includes helmets, the tunics, the corselets, the swords, the targets, and the spears are like those of Greek mercenaries in Persian service in the fourth century. Velikovsky listed additional similar details.

Some of the art at the time of Ramses III appears to be taken from art of the time of Assurbanipal. The problem is that Assurbanipal invaded Egypt around 663 BCE. If Ramses III were later, it is clear artists could have borrowed from the time of Assurbanipal. Ramses III built a number of magnificent buildings. In the seventh century, Assurbanipal destroyed many buildings in Egypt. The temples of Ramses III survived, but the temples of "later" dynasties were destroyed. Velikovsky said, "It is a different matter if the dating of Ramses III is drastically revised. In order to judge the age of the surviving buildings of Ramses III on their own merits, they should be compared with those of the Hellenistic age in Egypt." (Part I, Ch. 4, »Temple Architecture and Religious Art«)

The entire period of the descendents of Ramses III occupying the throne is supposed to have been only two generations, although there were Ramses IV to XI. The dynasty is thought to have ended in unknown circumstances. Velikovsky said that with Ramses III identified as Nectanebo I, there should be clues about those who followed Ramses III. Velikovsky said, however, that those identifications would be more speculative because the latter rulers did not have as much documentation. He presents some of the information and suggests the possibility that Ramses IV may be Tachos and Ramses VI may be Nectanebo II, and "King Psammetich" may be Psamshek, governor under Arsames. Psamshek and Nekth-nebef are identified with a pharaoh of the seventh century and with a pharaoh of the fourth century, but both appear to belong to the fifth century.

The conventional time of the Twenty-first Dynasty is the time in Israel of the later Judges and Kings Saul, David, and Solomon. The histories of Israel and of Egypt are interwoven through all of those people. *Ages in Chaos* identified the Egyptian rulers and events of the time. Velikovsky reviewed that information for those who had not read the other book.

What conventional chronology calls the Twenty-first Dynasty is said to be from about 1100 BCE to about 945 BCE. The documents seem to describe what appears to be a group who preceded, was contemporaneous with, and followed in time the Twentieth Dynasty. "The Twenty-first Dynasty is a succession of hereditary priest-princes who resided in Thebes, in Tanis, but mainly in the oases of the Libyan desert – el-Khargeh, the southern oasis, and Siwa, the northern." (Part II, Ch. 1, »A Chimerical Millennium«). Part II of *Peoples of the Sea* describes details of trying to reconstruct that period.

The papyrus of Ourmai is dated to the early Twenty-first Dynasty. He described details of invaders controlling Egypt. No invaders are known in the conventional chronology. Velikovsky presents evidence indicating the details are similar to the descriptions of Cambyses' conquest of Egypt in about 546 BCE.

Another papyrus is dated to the Twenty-first Dynasty but a few generations later than that of Ourmai. The papyrus describes a trip by Wenamon to Lebanon. This document has a number of clues that indicate it was probably written much later. The Egyptian papyrus contained a number of Hebrew words. One translator drew attention to the fact that a late Hebrew source contains a reference to the same shipping company mentioned by Wenamon. The translator thought it unusual that the same company had ships navigating along the Syrian coast for over nine-hundred years. A supposedly later Wenamon erected a shrine to one of the gods during the time of Nectanebo II. There is support for the opinion that the Wenamons were the same person. During the last ten years of the last Persian domination over Egypt, a dignitary or curate of the temple of the god Thoth at Hermopolis was called Petrosiris. In his tomb, an epitaph designates the Persians as "foreigners" who "had come and invaded Egypt." The same words are used in the Ourmai papyrus from the Twenty-first dynasty.

Velikovsky noted that placed in the time of the conventional chronology the Twenty-first Dynasty has no synchronous points of contact between Egypt and its neighboring countries. If the Dynasty is placed in the Persian period, there are numerous points of correspondence with Egypt and its neighboring countries. I have listed only a few examples from Velikovsky's extensive discussion about this and other periods. Since I could only mention a small amount of the material, nothing I presented was meant as proof. I tried to present an idea of what he concluded and a few of the reasons why. I may have misinterpreted some of the areas, and any of the mistakes are mine. To have a more detailed understanding of the reconstruction, you should read the original books.

Overview of *Oedipus and Akhnaton*

In 1960, Velikovsky published *Oedipus and Akhnaton*, which was based on research that he did when he first went to the United States in 1939. That work had been set aside when he discovered the clues that eventually led to the publication of *Worlds in Collision* and *Ages in Chaos*. In *Oedipus and Akhnaton*, Velikovsky noted strong similarities between the legendary character Oedipus and the Egyptian Pharaoh Akhnaton.

This work is almost totally independent of Velikovsky's other books. There is one major connection to *Ages in Chaos*. If the reconstruction of Egyptian history, as presented in *Ages in Chaos* (and presented earlier in Velikovsky's *Theses for the Reconstruction of Ancient History*), is basically sound, then the Oedipus legend could have sprung up in Greece soon after the events occurred in the life of the Egyptian Pharaoh Akhnaton. Otherwise, there is about a five-hundred year delay between the life of Akhnaton and the popularization of the Oedipus legend in Greece. Since the Oedipus legend is mentioned in the Odyssey, the story can be placed before the seventh century BCE.

Akhnaton is conventionally treated as the first monotheist. Velikovsky noted that, based on details of the life of Akhnaton, he was probably not a monotheist at all. Velikovsky also noted that if the revised chronology is correct, even if Akhnaton were a monotheist, he lived a number of centuries after the earliest known examples of monotheism.

With only these tangential connections to his other works, Velikovsky's *Oedipus and Akhnaton* was largely ignored by scholars, although a leading classicist, Professor Gertrude E. Smith of the University of Chicago wrote a favorable review of the book. Not everyone agreed. W. F. Albright, a previous opponent of *Ages in Chaos*, opposed the conclusions presented in *Oedipus and Akhnaton* based on the improbability of a cultural exchange between Egypt and Greece at such an early time. Albright held that view in spite of the fact that Mycenaean ware was found in abundance in the capital city of Akhnaton, and a seal bearing the name of Akhnaton's mother turned up in a Mycenaean grave in Greece.

The Legend

A number of treatments of the Oedipus theme exist. Sophocles presented three plays: *Oedipus Rex*, *Oedipus and Colonus* and *Antigone*. Aeschylus also wrote three plays, of which only *The Seven Against Thebes* remains. The names and details change, but the major plot is recognizable in each. It generally proceeds along the following lines: A royal couple has a baby son and, wishing to know what great things are in store for the child, consult a local blind soothsayer. Unfortunately, instead of the usual niceties provided by fortune tellers, the royal couple is told that the son will grow up and kill his father, marry his mother and corruption will ruin the kingdom. Disturbed by this prophecy, the couple decides to allow the child to die by abandoning him in the wilderness. He is rescued, however, and taken to live with a royal family of another country. Very little is known about his childhood.

When Oedipus became a young man, there was gossip around the royal residence which made him wonder about his past and future. As was the custom, he went to a soothsayer to ascertain his destiny and was told that he would kill his father, marry his mother, and bring disaster to the kingdom. Since he loved the people he thought were his parents, he decided to leave the country. While traveling, he encountered a man with whom he argued and killed. As it happened, this was his real father. Oedipus then ventured to the city of Thebes outside of which stood a Sphinx who desired human sacrifice. The Sphinx asked Oedipus a riddle, and when Oedipus correctly answered it, the Sphinx killed herself. The people of the city were so happy about the demise of the Sphinx that they offered to let Oedipus marry their recently widowed queen. He accepted, although she was a bit older than he.

The country started to decay and the people began to suspect some cause related to the morality of the monarchy. An investigation revealed that the king was living in not-so-holy matrimony with the queen, who was none other than his mother. When she discovered this, she committed suicide. Oedipus blinded himself, lived in seclusion for a while, and then went into exile.

There is also a side plot where an ambitious uncle helps to increase the unpopularity of Oedipus so his sons will become rulers. The uncle then causes the sons to fight. After they kill each other, the uncle marries the wife of one of the sons. She was also the dead husband's half-sister. The uncle then becomes the ruler. It sounds like a typical day-time television story, yet it was written over twenty-five hundred years ago.

Correlations to Egyptian History

While reviewing the life of Akhnaton, Velikovsky noticed a number of striking parallels between the life of Akhnaton and that of the Greek figure called Oedipus. In addition, several items of the Greek legend were of Egyptian origin and, theoretically, should not have been included in a story created in Greece by Greeks.

The following material summarizes some of the major points of similarity between the lives of both personalities. Additionally, a few of Velikovsky's intermediate conclusions are given. This outline is intended only as a guide to the book and is not intended as a proof. For extremely well-documented material presented in a fascinating manner, the reader is referred to *Oedipus and Akhnaton*.

Two of the lines on Akhnaton's genealogy are not always found in standard descriptions of Akhnatons's life. One of these is the line connecting Beketaten, the daughter of Queen Tiy, with Akhnaton. It is known that Queen Tiy was Akhnaton's mother, so under today's mores it is difficult to consider Beketaten as a by-product of the union of these two. Some people prefer to think she was the daughter of Amenhotep III, Akhnaton's father. However, records indicate that she was born after Amenhotep III had been dead for about six years.

Thebes

Most of the action in the Oedipus story took place in Thebes. A Greek, Boeotian city was called the "seven-gated" Thebes because of its outer wall with seven gates, and in order to distinguish it from the "hundred-gated" Thebes in Egypt. The legendary creature that watched over Thebes in Boetia was not a familiar Greek mythical figure. It was the Sphinx and called that by the Greek tragedians. The Sphinx, however, originated in Egypt. From actual images preserved in relief, it is known that there was a historical Egyptian Theban Sphinx to whom human sacrifices were made in the 18th Dynasty.

Soothsayer

Amenhotep, son of Hapu, no known relationship to Amenhotep III or Amenhotep IV, was considered extremely brilliant and able to see the future. He was renowned as a soothsayer and was frequently consulted by royalty. Tiresias, of the Oedipus legend, possessed the same characteristics as Amenhotep, son of Hapu. Both were blind as well.

The Son

Queen Tiy, comparable to Jocasta of the legend, had a son of whom nothing is known until he claimed the throne after his father's death. The son, Akhnaton, appears to have had swollen thighs, since the court artists depicted him in such a way as to emphasize this particular characteristic. Interestingly enough, Oedipus in Greek means a swelling of the foot or a swelling of the leg.

When Akhnaton was a young ruler, he used the epithet "Who lived long" or "Who survived to live long". This is possibly a result of his having survived some crucial event in childhood which might have caused him to consider any additional life as "living long". This event may have been similar to the attempted killing of Oedipus. Akhnaton also called himself son of the sun. Some ancient sources have Oedipus change his parentage from Laius to Helios.

Killing the Father

Akhnaton (Amenhotep IV) erased the name of Amenhotep III from the various monuments on which the name was inscribed. Erasing the name or memory of a person meant eliminating that person forever in the spiritual world; hence, Akhnaton "killed" his father.

After this act, Akhnaton instigated new religious practices. Theses actions earned Akhnaton the title of the "first monotheist." He probably was not a monotheist at all, let alone the first monotheist. If the revised chronology is correct, Akhnaton lived several hundred years after other known monotheists.

Part of the reasoning which led to the conclusion that he was monotheistic was that he changed his name from Amenhotep IV which contained the name of a god he wanted to eliminate from the religion. More significantly, he erased his father's name which also contained the name of the god, Amen. It is thought that, if he went to all the trouble to chisel out the name, he was serious about stopping the worship of this god. However, Velikovsky suggested that, since Akhnaton did not remove the name of Amenhotep II, Akhanton's interest was not only in religion, but in "killing" his father for eternity.

Marriage

Mitanni, where Akhnaton probably lived as a child, had Iranian customs which considered mother-son marriages holy. Akhnaton used the epithet "Living in Truth" possibly because he openly portrayed this relationship and tried unsuccessfully to have it accepted by the Egyptians in general. This added to the dissatisfaction that later helped in the removal of Akhnaton. There were other indications that Akhnaton was knowingly married to his mother. Akhnaton insisted that he was "husband of his mother, and his mother Queen Tiy was called "King's Mother and Great Royal Wife." Under today's customs, translators find this phrase incomprehensible.

Akhnaton apparently was polygamous. In addition to the implied marriage to his mother, Queen Tiy, he had some younger wives. One of these was the well-known Nefretete. Akhnaton, Nefretete and their offspring are often pictured facing Queen Tiy and Beketaten. Tiy and Nefretete did not appreciate each other and the conflict seems to have eventually caused a reduction of Nefretete's power. Oedipus had the same problem. His mother-wife Jocasta was not overly fond of Euryganeia, who was referred to as his "younger wife." Jocasta also had the younger wife's influence decreased or eliminated.

Ruin

Problems with other kings and famine made the Egyptians think they were being punished because of the acts of Akhnaton. Details of some of the trouble can be found in the el-Amarna letters (see *Ages in Chaos*). The displeasure of the people was also significant in the Oedipus legend.

Length of Rule

Ay, Brother of Tiy, advertised the problems and encouraged revolt against Akhnaton, as did Creon against Oedipus. Akhnaton possibly lived for a while in seclusion at a minor palace located near the main palace. Depending on the source, both Akhnaton and Oedipus were credited with reigns of both seventeen years and twenty years. The time in seclusion may have been three years which was probably not counted as actual time of reign.

Blindness

Akhnaton and Oedipus became blind in their old age. Oedipus blinded himself, but Akhnaton possibly was blinded by the same disease that caused his enlarged thighs.

Suicide

In the legend, Jocasta committed suicide. The mummy of Queen Tiy was identified as such only in October of 1976. The news report from the University of Michigan discussed only the identification procedure and did not mention a possible cause of death.

The Brothers

After Ay helped to depose Akhnaton, Ay may have encouraged Smenkhkare and Tutankhamen, two of Akhnaton's sons, to alternate their rule. In the legend, Creon created a similar situation with Polynices and Eteocles. Ay and Creon then each encourage Tutankhamen and Eteocles, respectively, to retain the throne; this incited Smenkhkare and Polynices, who each acquired armies and attempted to regain power. In the ensuing battles, real and supposedly legendary, all four characters were killed. Ay then married the widow of Tutankhamen since the power in Egypt was acquired through the female blood line. In the Greek version, Creon married the widow of Eteocles. Both Ay and Creon had also had previous wives who died young. Eventually Ay and Creon were both dishonored after their own deaths.

The Burials

In the legend, Creon decreed that Polynices should not be buried, but that the great hero and defender of the land Eteocles, should have the most lavish burial possible. In real life, Ay made the burial of Tutankhamen an unforgettable event. In fact, about the only thing King Tut is noted for is having an extravagant funeral.

The sister of Polynices, Antigone, buried the body of Polynices, and for her efforts Creon had her entombed without benefit of death. When Creon finally decided to free her, she had already hanged herself with her scarf. Some evidence indicates a hastily buried body near Tiy's tomb may have been the body of Smenkhkare. Not far from this tomb was a pit with evidence that perhaps someone had been entombed alive. Also found was a scarf with the name that Smenkhkares' wife called him. No remains of a body were found.

Miscellaneous

The rock tombs used in the legend were not common to Greece, and the Greeks did not place as much importance on burial methods as did the Egyptians.

In the legend, Antigone performed some mutually exclusive acts, such as going into exile with her father and then being buried alive near the tomb of her brother. In the life of Akhnaton, Beketaten may have gone into exile with her father, while Meritaten was probably entombed near Smenkhakare.

When Velikovsky published *Oedipus and Akhnaton*, there was some question about a few of the relationships discussed. For example, it was not clear that Tutankhamen and Smenkhkare were brothers. However, later investigations supported that identification. Harrison, Abdalla, and Connely performed analysis or the remains of Tutankhamen and Smenkhkare. X-rays of the skulls showed that they are the same basic shape. Special blood tests were made, and one conclusion was that Tutankhamen and Smenkhkare had the same parentage. Also King Tut probably lived to be only nineteen years old and did not die from tuberculosis as previously thought. Although an ancient painting does depict Tut in battle, X-rays indicate that he did not die from a blow to the head.

Harrison et al also felt that these two brothers and Akhnaton (Amenhotep IV) may have been sons of Amenhotep III. However, Dr. I. E. S. Edwards, Keeper of Egyptian Antiquities at the British Museum, took a more non-committal approach by saying that Amenhotep III was either Tutankhamen's grandfather or his father. That Amenhotep III might have been Tut's grandfather is more in line with the chronological analysis given by Velikovsky.

Professor Lewis M. Greenberg has also presented a new dimension regarding the religious innovations of Akhnaton. Professor Greenberg suggested the possibility that the worship of the god Aten may have been the result of a synthesis of Egypt's solar theology with a new cosmological phenomenon of considerable importance. This referred to the worship of Venus. Greenberg refers to the conflicts between the religious leaders and Akhnaton, partly due to the Pharaoh's novel marriage arrangements, and also points out the probable needs of the new religion. Greenberg then poses some questions about the new religion that would arise if it were merely a revised form of sun worship. For example: "Why was Aten represented with rays emanating in arcuated fashion from one side only, as a comet's tail, as opposed to the standard portrayal of the suns rays in a 360 degree sweep?" Also, "Why does the Hymn to Aten say 'it rises like the living Sun' when one would not normally say that the sun rises like the sun?"

It may be that a similar legend originated earlier than the time of Akhnaton and interest in the story was revived after a real life version occurred in the Egyptian royal family. In either case, several people agree that Velikovsky made a strong case that the Greek version has details inspired by the Egyptian events. For example, when Dr. Cyrus Gordon was Chairman of the Department of Mediterranean Studies at Brandeis University, he wrote an article about *Oedipus and Akhnaton*. He concluded that Velikovsky made a strong case for the correlation of the events in the lives of the real Akhnaton and legendary Oedipus. Dr. Cyril D. Darlington, at the University of Oxford, agreed that Velikovsky demonstrated that the Greek version of the Oedipus legend had an Egyptian origin.

Overview of *Mankind in Amnesia*

Mankind in Amnesia is an extremely interesting book and should be fascinating reading for people who are only slightly familiar with Velikovsky's other works. He coined the term "Collective Amnesia" in 1950 in *Worlds in Collision*. The basic theme of *Mankind in Amnesia* is collective amnesia. The idea is described largely in two sentences in *Worlds in Collision*. He said, "The memory of the cataclysms was erased, not because of lack of written traditions, but because of some characteristic process that later caused entire nations, together with their literate men, to read into theses traditions allegories or metaphors where actually cosmic disturbances were clearly described." Also, "Disaster may come, not from another planetary collision, but from the handiwork of man himself, a victim of amnesia, in possession of thermonuclear weapons."

Professor Eugen Bleuler, prominent Swiss psychiatrist, was the dean of psychiatry for several decades including the 1930's. He is best known for his introduction of the term schizophrenia. Bleuler wrote an introduction to a 1930 paper by Velikovsky containing basics of the idea of a collective mind. Bleuler said, "The ideas of the author [Dr. Velikovsky] appear to me very much worthy of consideration."

Amnesia affects long-term memory. Various subsets of amnesia include memory loss caused by psychological trauma, close association with a stressful event that involves serious threat to life or health, and long-term repressed memory that is the result of psychological or emotional trauma.

Velikovsky detailed five components of the collective amnesia concept in *Mankind in Amnesia*. These components are 1) subconscious memory of devastating events in the past can be transferred from generation to generation, 2) people try to keep the memory alive by referring to the events symbolically, 3) people ignore or do not recognize the data as indicating a traumatic event, 4) people may act irrationally when confronted with the realization that ancient catastrophes occurred, and 5) people try to relive the events often wanting the bad to happen to someone else.

An important term with respect to collective amnesia and to item 1) is "mneme". Mneme (now spelled "meme") consists of a self-propagating unit of cultural evolution having an analogous resemblance to the gene (the unit of genetic information). This is basically the type of transfer envisioned by Velikovsky with respect to collective amnesia.

Richard Dawkins DSc, FRS, FRSL is an eminent British ethologist, evolutionary theorist, and popular science writer who holds the Charles Simonyi Chair in the Public Understanding of Science at Oxford University. He used the term "meme" in the previous context in 1976. He believes he coined the term and the concept; however, John Laurent, School of Science, Griffith University said the concept "appears in the book *The Soul of the White Ant*, by the Belgian dramatist, essayist and amateur entomologist, Maurice Maeterlinck, first published in 1927." Laurent said it is difficult to believe that Dawkins was unaware of Maeterlinck's work. Velikovsky, in *Mankind in Amnesia*, referenced Jung as using the name "mneme" and the concept for collective unconscious around 1921. Jung said that the "collective unconscious does not develop individually, but is inherited. Jung referenced Semon. It appears that Richard Wolfgang Semon, a German zoologist and evolutionary biologist, was the first to use the word in the context of psychological memory transfer from generation to generation. He wrote *Die Mneme* (1904), translated into English as *The Mneme* (1921), and *Die mnemischen Empfindungen* (1909), translated as *Mnemic Psychology* (1923).

Velikovsky said, "The existence of a racial memory does not mean that an impression absorbed by one generation can be remembered by the following ones, but that impressions, especially traumatic and repetitive impressions, experienced by man of the forebears, may become a permanent though unconscious mneme or mneme complex, providing adequate responses in suitable situations."

No matter who first used the word "mneme" or the idea about generational transfer of psychological impressions, Velikovsky was the first to use the term "collective amnesia" in that context. Not only does the work of Dawkins support that idea, but numerous others seem to have accepted the idea. If you do a search for "collective amnesia" on the web, you obtain hundreds of results. The phrase is used in the same way that Velikovsky used the phrase. Listed below are a few examples:

- *Collective Amnesia: Japan's Crusade to Forget*
- *Collective Amnesia: Vietnam War*
- *Beyond Collective Amnesia: A Korean War Retrospective*
- *The cost of collective amnesia (slave trade in Africa)*
- *The world's collective amnesia* by Joseph Farah
- *Collective Amnesia of the Jewish Holocaust in Romania and Current Narratives of National Identities* by Alexandru Cuc, M.A.
- *Collective Memory, Collective Amnesia. The struggle for memory in post-authoritarian Argentina, 1991-2005*

Jung said that the collective unconscious "consists of pre-existent forms, the archetypes". Velikovsky said, "What made these forms populate man's mind was never explained by Jung." Velikovsky also said, "How did some true archetypes enter the human mind and settle there to plague it, and to pass from generation to generation? – was not even asked."

Velikovsky did note that Jung realized in myth there is an "unconscious psychic process" at work. Jung further stated that the work of mythologists is futile when they use solar, lunar, meteorological and similar ideas to explain myth. He said that mythologists 'absolutely refused' to see that mythological images and contents reflect deep, ingrained psychic phenomena.

In discussing item 2) Velikovsky said, "But simultaneously with the phenomenon of incipient, almost willful, amnesia, one can observe an opposite current, a conscious effort to preserve the memory of events that shook the framework of the earth, events in which the entirety of nature – sea and land, Sun and Moon, and all the celestial host – participated." He later noted that, "It is therefore not surprising to find in Plato a number of passages dealing with the subject of global or even cosmic upheavals."

The following is an example of 3) in the components listed previously. Velikovsky described the psychological equivalent of scotoma in ophthalmology. He said, "Psychological scotoma is an inability to observe certain phenomena or to recognize certain situations though they are obvious to other persons."

In 1993 in *The New Catastrophism, The Importance of the Rare Event in Geological History,* geologist Derek Ager said, "Even within the brief life of mankind (with 99% of it in the 'Stone Age') there were great geological events that are not recorded in our histories" [p. xviii]. As Velikovsky demonstrated, the events were recorded. The recorded data is called myth and is still ignored by some researches such as Ager. Part of the basis for ignoring the data is associated with what Velikovsky called "collective amnesia".

Ager was not the first to overlook the obvious. Plato said, "At any rate they seem to have been strangely forgetful of the catastrophe." (Plato, *Laws* iii, translated by R. Bury.)

In *Worlds in Collision*, Velikovsky said, "It is a psychological phenomenon in the life of individuals as well as whole nations that the most terrifying events of the past may be forgotten or displaced into the subconscious mind. As if obliterated are impressions that should be unforgettable. To uncover their vestiges and their distorted equivalents in the physical life of peoples is a task not unlike that of overcoming amnesia in a single person."

Velikovsky noted, "It has been immensely difficult for the mind of man to part with the conviction that his Earth is immovable and in the center of Creation, thus in the center of the system to which the Sun, the Moon and the planets belong, and in which the stars are without a clear purpose or design." By the time of Aristotle, the unchanging Solar System was ingrained. "The Aristotelian negation of the traumas of the past, built into a philosophical system that covers many fields of human knowledge, became the rock on which the Alexandrian schools of physics, geometry and astronomy of Archimedes, Euclid and Claudius Ptolemy were built." Velikovsky said, "When Galileo embraced the Copernican doctrine of the Earth and other planets revolving around the Sun, he broke not with the Scriptures, but with Aristotle." This desire to ignore the past events extended to modern times. "The success of Darwin, the speedy acceptance of his theory by academia, and the penetration of his theory into all things spiritual and material of the last hundred years was due to his assurance that the frame of this globe had never been shaken."

"Obviously there was a psychological need in Darwin to shut his eyes to contrary evidence."

Part of the problem in wanting to ignore the past was that "A moving earth is a less secure place than an unmovable one. Moreover, the system denied man the central role in the universe; this was injurious to his ego. It was also in conflict with the tenets of the Christian Church; did Jesus come merely to a very secondary planet, one of many?"

This leads to part of the discussion about item 4). During the initial discussion about Darwin, "None of these contradictions between his diary observations and his views in his major work was raised by any of his opponents. It was not brought up against him that he had no academic position in a university, or that his only scholastic degree was that of a bachelor of theology, or that he omitted all footnotes to his sources, so that it was often impossible for a reader to check on the data; none of these shortcomings was ever mentioned by his critics." Many of these objections were brought up when Velikovsky proposed his ideas about ancient catastrophes, although many of his ideas are accepted without credit going to him.

"The adherence to the dogma of uniformitarianism is a symptom of an all-embracing fear of facing the past, even the historically documented experiences of our progenitors, as recent as four score generations ago." (Ch. 2, »Two Forms of Fear«)

"The irrationally emotional reaction to *Worlds in Collision* of so many people, especially among scientists, it seemed to me [Velikovsky] must surely be caused by a hidden fear of knowing the events of the past, more than by an aversion to challenging the conventional notions of science."

"The reaction against efforts to bring to the surface of consciousness repressed contents that struggle to stay repressed can be violent and cause an outburst of hatred; the person trying to help another to bring up the suppressed may himself be accused of fomenting hatred and discord." That this is true is well known to anyone familiar with the Velikovsky Affair.

With respect to 5), Velikovsky said, "Organized warfare has its inspiration in the same terror. As the ancient Assyrian kings went to war, they compared the destructiveness of their acts to the devastations caused by the astral deities at the time of the upheavals."

Velikovsky gave a historical chronology of groups expecting the end of world soon. He also quoted St. Clair Drake, a professor of anthropology and sociology at Stanford University who described "the phenomenon of what I call 'The Samson Complex' – the passionate desire to create the catastrophe that will destroy both them and their oppressors."

St. Clair Drake also said about Seventh Day Adventists, "The Adventists I knew were *not* looking forward to possible atomic annihilation of mankind with dread and horror, but with hopeful anticipation. They believe that this was the mechanism by which the new earth prophesied for the Millennial Age would come into being." St. Clair Drake said about Adventists and Jehovah's Witnesses, "How do we explain their acceptance of, and even welcoming of, disasters that will destroy millions, while in their personal lives they stress pacifist values?"

In part of the book, Velikovsky gives the historical chronology of people trying to cope with psychological problems caused by the ancient catastrophes and how literature repeats the stories in symbolic form. This history is very entertaining even if you have not read his other works.

Velikovsky also said that a number of writings describe Exodus event activities as if the events would occur in the future. The book of Revelations is one of those. Another is the poem *Darkness* by Lord Byron. "Byron showed no awareness of the similarity in his poem to conditions at the time of the Exodus and the wandering in the desert, when he spread before his readers this picture of a world coming to an end." Also, "Lord Byron did not write an eschatological picture of a dying world of the future, as he thought he did: he drew from a common spring of all men, even of all the animal kingdom."

As noted in the list of descriptions of collective amnesia found on the web, some modern experiences give similar psychological effects as the ancient catastrophes. Velikovsky mentioned a book about Hiroshima. He said, "What fixed my attention on Paul Goodman's essay was his insight in comparing the experience of the persistent reaction of the survivors of Hiroshima with the experience and the subsequent attitude of the Israelites of the days of the Exodus, an event long passed into history; he actually starts his piece by referring to Martin Buber's book on Moses: '… the bible story cannot be taken literally yet is not unhistorical: something happened that was, to those people, super-natural or crazy, and the account we have received was their attempt to cope with the experience, to regain their wits, to reconstitute themselves in the world that had been transformed.' "

The Harvard philosopher George Santayana said, "Those who do not remember the past are condemned to repeat it." Oddly, the saying was on the "throne" of Jim Jones at the infamous Jonestown. Jones had the followers in his cult commit mass suicide. Those who thought that was too big of a catastrophe for them were shot. Velikovsky described the psychological process involved in that and similar events.

Velikovsky quoted Norman Cohn from *The Pursuit of the Millennium* which described a sect in Germany in the fifteenth century that was similar to the cult of Jonestown. 'These people were convinced that as they beat themselves an angel named – surprisingly – Venus watched over them. Their skins all red with blood seemed garments for a wedding-feast, the skirts which they wore during flagellation they called robes of innocence.' Velikovsky noted that it would be no surprise to people who read *Worlds in Collision* why the angel was named Venus.

As an aside, Velikovsky said, "The five-pointed star – the ancient symbol of the planet Venus – adorns the headgear of every American, Soviet, and Chinese soldier."

Velikovsky's Solar System

(by Lewis M. Greenberg)

By studying the records of antiquity – ancient texts, astronomical data, various religions, myths, and rituals – Immanuel Velikovsky was able to propose a novel and revolutionary cosmological reconstruction of the recent history of our solar system. He concluded, among other things, that Venus was a recent entrant to the system's inner planetary arrangement and the current configuration did not coalesce until historical times, shortly after the beginning of the seventh century BC.

From his theoretical construct, Velikovsky made a number of advance claims or predictions. Before presenting a few of these, it must be emphasized that the research he performed and the conclusions he reached occurred more than a decade prior to the beginning of the Space Age. Additionally, there were no personal computers, no Internet, xeroxing, fax machines, TV, or reasonable telephone lines available at that time. Just imagine trying to conduct research today under those conditions while trying to simultaneously formulate and substantiate what we would now term cutting-edge astronomical theories.

Based upon his research, Velikovsky offered a series of claims:

Venus

Venus, due to its recent birth (from the planet Jupiter) – perhaps some 4000 years ago – and violently cataclysmic history should display certain primordial conditions. Thus, it must still be hot under its cloud cover to such an extent that it gives off heat having been in a state of candescence only 3500 years ago (*Worlds in Collision*, Part I, Ch. 3, »The Battle in the Sky«, Epilogue). This prediction was verified in early December of 1979 when Pioneer Venus 2 reached that planet and reaffirmed the findings of the earlier Mariner II probe of 1962 (see *KRONOS* IV:4, Summer 1979, pp. 1-15), Yet, the scientific community has chosen to quibble over the word "hot" and Carl Sagan even tried to disparage the use of the word "candescence", claiming it was not the same as "incandescence". Nevertheless, Webster's Dictionary considers them to be synonymous.

As it happens, Venus today has a surface temperature of some 900° F – enough to melt lead. An ad hoc theory attributing the source of its heat to a "greenhouse effect", a "runaway greenhouse effect", or an "enhanced runaway greenhouse effect" has been put forward by Sagan and others. All of the greenhouse effect theories have been repeatedly debunked over the past 40 years, but to no avail (see references below). The greenhouse theories are inadequate to account for Venus' high surface temperature, especially for that on its nighttime side.

Velikovsky's study of the ancient sources – particularly in regard to the time of the Hebrew Exodus from Egypt – also led him to posit that hydrocarbons and sulfur would be found in the Venusian atmosphere, and petroleum fires would be found on Venus itself. This conclusion was actually confirmed, for the most part, but never acknowledged by the Pioneer-Venus Early Findings report (see *KRONOS*, ibid., pp. 4-10; *KRONOS* IV:3, Spring 1979, pp. 63-66; also see I. Velikovsky: »Venus and Hydrocarbons«, *Pensée* IVR VI, Winter 1973-74, pp. 21-23).

References for the repudiation of the "Greenhouse Effect" theories:

See e.g. *Yale Scientific Magazine*, »Venus – A Youthful Planet« by Velikovsky which was first written in 1963 but only published in April of 1967 [reprinted in *KRONOS* IV:3, Spring 1979, pp. 61-63]; F. Jueneman, »pc«, *KRONOS* I:3, Fall 1975. pp. 79-80; R. Juergens: »Velikovsky and the Heat of Venus«, *KRONOS* I:4, Winter 1976, pp. 86-92; L. M. Greenberg: »The Venus 'Greenhouse Theory' – Debunked«, *KRONOS* III:2, Nov. 1977, pp. 132-134; L. M. Greenberg: »Velikovsky and Venus«, KRONOS IV:4, Summer 1979, pp. 10-12; S. Mage: »Back to the Drawing Board«, ibid., pp. 13-15; V. A. Firsoff: »On Some Problems of Venus«, *KRONOS* V:2, Winter 1980, pp. 64-65.

Jupiter

One major deduction from his study of ancient civilizations was Velikovsky's belief that the planet Jupiter probably emits radionoises. He presented his conclusion in a Forum Lecture Address at Princeton Univ. on October 14, 1953 (see *Earth in Upheaval*, Supplement, NY, 1955, p. 297). This proposal was confirmed on April 5, 1955. From his research, Velikovsky had concluded that Jupiter was not "an inert gravitational body" and sent out radionoises of a non-thermal origin (see *Earth in Upheaval*, p. 295 and I. Velikovsky: »On the Advance Claims of Jupiter's Radionoises«, *KRONOS* III:1, Fall 1977, pp. 27-30; *Worlds in Collision*, Part II, Ch. 4 »Synodos«).

At a McMaster University Symposium held in August of 1974 in Hamilton, Ontario, to discuss Velikovsky's theories, physicist James W. Warwick grudgingly admitted that "Velikovsky had a valid but intuitive inference to make ..." regarding Jupiter's radionoises. But Warwick went not further because he could not comprehend Velikovsky's Jovian prediction with respect to the latter's theories (see *Pensée* IVR VIII, Summer 1974, pp. 41-42). Evidently, Warwick could not see "the big picture" as it pertained to Velikovsky's cosmological scenario.

While Velikovsky had not publicly revealed that much about Jupiter at the time (a book on *Jupiter of the Thunderbolt* remains unpublished), he did allude to Jupiter as a "dark star" in *Worlds in Collision*, Part II, Ch. 9, »The End« and from his knowledge of Jupiter's activities in ancient times as well as his awareness of cosmic electromagnetic forces, Velikovsky was able to make a plausible deduction regarding Jupiter's "stellar characteristics". Warwick could not grasp this fundamental point despite his own awareness of the complexities of the radio spectrum.

A final rhetorical question by Warwick is of particular interest: "I who am a specialist in the field am moved to ask myself, 'Did this physician writing in 1954 know more about the physics of radio emissions from planets than this astrophysicist 20 years later?'" – to which one could reply with a resounding "YES!".

Mars

Velikovsky claimed that the surface of the planet Mars would reveal a catastrophic history resulting from its near encounters with Venus and Earth. The various Martian probes have certainly confirmed this. The discovery of water on Mars would also affirm one of the things that Velikovsky mentioned during his discussion of that planetary body. Velikovsky also predicted the presence of argon and neon in the atmosphere of Mars. In 1974 and, again, in 1976, Soviet and American Martian probes did reveal the presence of argon though the amount has been disputed (see *KRONOS* II:1, August 1976, pp. 105-109).

Saturn

Velikovsky's studies led him to conclude that Saturn was the source responsible for the Universal Flood or Deluge. He believed, therefore, that Saturn contains (or consists of) water and that the Saturnian rings also consist of ice. The latter has been found to be the case as a result of space probes. Velikovsky also expected Saturn to possess molecular chlorine and emit X-rays and/or cosmic rays. This still awaits confirmation one way or the other. Along with Jupiter, it is now known that Saturn is more "star-like" than planet-like. Both Velikovsky and Dwardu Cardona have presented material supporting the "stellar" aspects of both Saturn and Jupiter (see D. Cardona: »The Sun of Night«, *KRONOS* III:1, Fall 1977, pp. 31-36; I. Velikovsky: »Khima and Kesil«, *KRONOS* III:4, Summer 1978, pp. 19-23; D. Cardona: »The Mystery of the Pleiades«, *KRONOS* III:4, ibid., pp. 24-44; I. Velikovsky: »On Saturn and the Flood«, *KRONOS* V:1, Fall 1979, pp. 3-11; L. E. Rose: »Variations on a Theme of Philolaos«, ibid., pp. 24ff). [Velikovsky's opus *Saturn and the Flood* remains unpublished.]

Lewis M. Greenberg
Professor Emeritus
Moore College of Art & Design
Formerly Editor-in-Chief of *KRONOS*

Psychoanalytic Lecture

Psychoanalytic Lecture

The following is a transcript of a lecture by Immanuel Velikovsky as a guest speaker at a psychoanalytic class I taught in New York on April 5, 1979 – seven months before he died on November 17, 1979.

I assume you wish to hear from me of my meetings with some originators of psychoanalytic and psychological treatment and I will start with this theme.

I was the first time in Vienna before World War I and I was myself still gymnasium student. So, at that time, the name of Freud was not familiar to me. I was on my way to Palestine, today Israel.

My real first contact with Freud was in 1921, when I left Russia for good, just having accomplished my state examination for medical degree. The next two or three years I spent in Berlin. I was in Berlin for several purposes. One purpose, not really pursuit of science on my own, was in bringing together Jewish scientists and scholars to prepare the spiritual basis for creation of a Hebrew University. There was no Hebrew University in Jerusalem at that time. There was only a plot of land and a stone of dedication but nothing beyond.

The idea was to bring in publications, called *Scripta Universitatis* (this is half of the name, it was published also in Hebrew), the works of prominent Jewish scientists and scholars in various lands all around the globe. And of course, we traveled, I and Professor Loewe who was helping me in this plan. He was librarian of University of Berlin, an old Zionist. And we approached also Freud. Freud refused to participate, but his answer was very friendly. He said that his followers would not find a paper of his if it was published somewhere outside of the regular publication of the psychoanalytical society then.

My own interest in psychoanalysis and generally in psychological subjects, goes to the time, I would say, in 1928 when my mother died and I was of age 33. Some patients that I had in my care made me interested. I was in charge of their labor union, their insurance for patients. A wife of a man who was invalid himself, was treated for heart disease. After observing one of her – well so-called – heart attacks, I came to the conclusion that this was nothing else but hysteria. And that she was a perfect case for hypnosis.

Hypnosis, I observed in the University of Moscow, performed by Professor Popov. This usual thing that a woman – or man – could not lift his hand, or could not remember his name or there would be an illusion that she drinks wine, and it was not wine and so on. But this was already a number of years before. And I let this hysterical patient, presented it to the doctors, to the medical colleagues in Haifa and later in Tel-Aviv.

Next was my effort to penetrate this field by means of reading. Actually, my first paper, not the second one dealing with other subjects, psychological, I would say, or if you wish, dealing with extrasensorial perceptions to some extent, was prepared when I visited Zurich in 1930. I met Professor Bleuler, Eugene Bleuler. Eugene Bleuler, you should know, was – I would say – the leader in psychiatric research, was the most dominant figure in the world in this field, and Carl Jung was one of his assistants. He put Carl Jung to study Freud. And actually, it was after Professor Bleuler became interested in the work of Freud that doors of academia opened to Freud. I showed him my paper on *Physikalische Existenz der Gedankenwelt*. He wrote me a beautiful preface – not immediately, after reading a second time. The paper would have not been published, if it had not had the preface of Eugene Bleuler. Even so, I had first one refusal, rejection. On the second try it was accepted in one of the most prestigious magazines in the field of neurology and psychiatry. I sent the paper to Freud. He wrote me, similarly like Bleuler, that he himself had similar ideas, almost identical he said, ideas which he didn't yet publish.

And so when I was in 1930 in Zurich, I made an effort – at least a try – to see whether the teaching of Jung would be in harmony with my own feelings and approaches. So I visited him, Carl Jung. He was a tall man, permanently with his Danish dog, "Fauso" – very big. And all went well until I mentioned that he was a pupil of Freud. He became bristled! He didn't like the reference to himself as a pupil of Freud. Nevertheless, I suggested that we should spend some time together. He recommended me a close (to him, in every sense close) analyst, a lady. I visited her. On the second appointment, it was actually first appointment after introduction first time, I decided not to continue. Also, coming from Palestine where I had family, two children and wife, I could not go away for very long, even having prepared myself. And even in those days, analysis did not take so long as today for preparation.

In 1933 I went to Vienna. I selected in advance to see Dr. Stekel. In a book that I had, that was describing various forms of psychotherapy and psychoanalysis, I thought that in this case the shortest preparation and a short analysis would be the way that would suit me best, also because of my temperament. I was feeling that I would not be able to sit hours behind the couch and wait for a patient to open his mouth. And then let him go pay the hourly fees.

But I did not come to Stekel empty handed. Stekel wrote many volumes. His work on neurosis, I possess them all somewhere if my archive in Tel-Aviv still exists – there is a room or two with this archive, but what left, I don't know – but there were books of Stekel and he wrote many, about 12 or more, a series on Zwangsneurose, or neurosis of obsession, Angstneurose, neurosis of fear and so on. The first volume of this series in German dealt with the story of a patient, a lady patient, opera singer, and described this analysis on many pages, I don't remember now how many, maybe fifty, maybe a hundred, with all her, what she brought us in free associations and the dreams that she dreamt and so on. I came from Russia so I am not very knowledgeable in German literature, but I observed immediately that what she was telling him, Stekel, was actually from Faust by Goethe. All the analysis was, all her dreams, all her quotations were from Faust of Goethe. So I wrote a reinterpretation of this analysis and with this I came from Palestine to Stekel, three years after I was in Zurich.

He read the paper. The fact that he was a magnanimous man, generous, was that he brought together his group, whatever it was, maybe as large as here, some 20 people or more, that were at that time in Vienna, preparing themselves for analysis, for being analysts. He brought them together and let me read the paper which was practically a devastating criticism of his way of analyzing this singer. Then he admitted that her name was Margaret. That he added himself. But this was in some respect also my undoing, because after a short time, a rather short time, he told me, "You are a master and I don't need to give you analysis. You can do it yourself."

This was the first generation of the Freudian pupils. You have to know that Freud started his work in 1895, the year I was born, and he had his nine years of splendid isolation. He had no followers and had no pupils and he felt best. In 1904 he had two pupils. These were Stekel and Federn. About Federn, I will speak later. I think in 1905 nobody joined; and only in 1906 he had more of them, and soon there was already a little association. Not many, a dozen people around in Europe, maybe two dozen, and maybe two or three in America.

Now, in 1900 Freud published his book *Traumdeutung*, the analysis of dreams, the best known book. In Vienna, psychoanalytical society had its monthly meetings and I was at the meeting. He was not present because he was going from one surgery to another. He had a cancer of the jaw and went through 16 or 17 surgeries altogether. But his daughter, Anna, was at the meeting, and this was the meeting when the new book of his appeared, a *Neue Folge* of the new series of *Introduction into Psychoanalysis*, and here there were the very same sentences that were in my book and he wrote to me at that time he has identical ideas, he even expressed them identically, which resulted later in a certain article published in New York but I will not go into this detail.

The fact was that the meeting was in an uproar. They did not wish to have anything in common with problems like telepathy or any kind of mediums or something of the kind. And so they would not revolt openly against Freud, but it was closest to a revolt that he would figure out. And only two in the meeting sided with Freud, myself and Paul Federn. Paul Federn was the chairman of the meeting, also chairman and president of the International Psychoanalytical Society in those years. Freud was member. He was the chairman. Now, like Stekel, he joined Freud in 1904. And as it usually is in Vienna, after the meeting, he invited me to join him. That was a beginning of a long friendship that continued till his death.

Freud himself I spent tête à tête, which means, one against one, on his birthday. It was not in Vienna, but in a suburb of Vienna where he went for this occasion and he left his guests or whoever he had on the terrace, and came to see me. We spent 30 or 40 minutes together. We had already corresponded long before. More than that, he already published several pieces of mine in *Imago*. So I was not foreigner to him; in *Imago* and *Psychoanalytische Bewegung*.

He made on me the impression of a fragile man. He was birthday 77. I have not written down what we discussed and regretfully I do not remember too well. I only felt what I read, then and later, too, that he was not a quick grasper of a man, of a character of a man. Even Ernest Jones, one of his pupils who wrote his biography in three volumes, he admitted, because of the experience that he had with various people, some of his friends and some of his opponents, that he was not a good "Menschenkenner", which means did not know to characterize or grasp a personality well. He was generally a slow thinker, in comparison to Stekel who wrote his book on dreams – well it was not pioneer work, the pioneer work was of Freud – but it was so much richer so much quicker, so much abundant in ideas. So, this was also one of the reasons why I went to Stekel.

But whereas at that time I thought, even maybe tried even to make a kind of a peace between Stekel and Freud, it is not clear why there was not a peace between them. Because Freud had parted with Jung because Jung was a mystic, and Freud was not. He parted with Alfred Adler; Alfred Adler was a socialist of a family of socialists, even of terrorists, and was ... all in his mind was struggle for power. And there was a complete rift between Adler and Freud, which could be observed in many places in Vienna. In Vienna analytical practice and theory occupied a very important place in that year of 1933. There were daily places where you could go and learn, some places where young people or schoolboys or others were advised in public, in open sessions by people of the school of Adler. Adler himself gave lectures in very large audience, auditorium maybe taking one thousand people. And Adler also had made a seminar in his own house, in his own apartment, a 30-day seminar and I took part, I and my wife took part. He was like a "Amokläufer". Just in one way he could not see left and right, it was only one way.

Wittels, the first biography of Freud which I read, himself a pupil of Stekel, expressed himself that the behavior of Freud against Stekel was like as if Stekel stole silverware from the table that he was invited. It could not be understood what was this rejecting of a man who was finishing his analysis three or four or five months. But he could do it, why not?

Among the letters that I exchanged with Freud was also one when I invited him to come to Palestine, and he wrote me, I believe it was in 1932 or earlier, and he wrote to me that because of his state of health he can exist only in his hospitality of his own house and with the care that is given him there, but nowhere else he would like to travel as much as to Palestine.

From 1924 to 1928 I practiced internal and general medicine in Jerusalem first. But from 1928, as I told you before, I started to intrude, I would say, without any diplomas on the walls, into the field of psychoanalysis. But I could remember from my university years in Moscow, well, I can say that the professor of psychiatry, Rosso Limo was highly regarded in Russia. And, actually, to his lectures not just students were coming, but many doctors from wherever they would, from Moscow, from outside of Moscow, would come and be present – something like the days of Charpot in Paris where Freud went in, I believe, in about 1890 or something. But I was surprised to read in a biography of

Leo Tolstoy, how Leo Tolstoy describes this Rosso Limo, whom he visited too, as a fool. So I don't now what really to say about all this.

And I had to take care of patients, neurotic patients and others because there was no analyst in Palestine. There was one psychiatrist in Jerusalem who knew me well from years when I worked in Jerusalem and patients were sent from all parts of the country.

Now, in 1939, I saw in a window of a bookstore the new book of Freud which was titled, *Moses and Monotheism*. Part of this work I read already before, namely in 1937, there was an international congress of psychologists in Paris. It is two years before World War II. And I went there and it was visited by a very great number of psychologists, also from America. And by chance it so happened that the paper that was read by the chairman of this convention congress, Professor Claparet from Switzerland, was on the same subject as papers that I prepared, namely about homosexuality and the effect of it in the war, the war actions with Turks, assassinating Armenians, and all this masculine and feminine purposes playing roles. With France represented with a small cockerel and its repeated confrontation with Germany. So I saw the first chapters of this new book already printed in *Imago* when I was in Paris and visited there the National Library. But here was the book. I purchased it in the window, the same time was another book, *Mein Kampf* of Hitler and I thought to whom I should give preference and occupy my mind first, and I selected the book of Freud.

And it did not take much time when I asked myself, "Is not one hero of Freud, which is practically the hero of this book, Pharaoh Akhnaton, the historical original of his other hero, Oedipus." Quite a few phenomena to whom I attention to this. In the next maybe two months I had already the plan of a book, *Freud and his Heroes*. And the time was before World War II and I figured out that if I will stay now in Palestine I will not be able to dedicate myself to scientific or scholarly or literary work, not because there were no more analysts – by that time from Western Europe came quite a number of them because of Hitler and Nazis opposing psychoanalysis, opposing Freud, and so on. Freud, as you know, had to finally – though he did not wish to – leave Vienna. (He lived actually in the same apartment for 47 years, from 1891 to 1938).

So I looked again into his dreams, dreams of Freud, this I performed once more what I did to Stekel. I reanalyzed Freud's dreams – namely, he gives associations, in his book of dreams there are about 16 or 17 dreams of his own, (and you had the chance to read it because Ruth, well, invaded my attic, my three rooms of my papers, found copies and distributed here – with a promise that she will bring them back). Now, from this you could learn that Freud carried in his subconscious, and maybe conscious mind, the idea to part with Jewishness. Ernest Jones, who wrote his biography, disagreed with me. He wrote in his biography that he spoke many times with Freud and knew that Freud was never contemplating this kind of desertion. Well, I answered Freud in my book on *Oedipus and Akhnaton* just in one passage.

Ruth: Happens to be right here. *Oedipus and Akhnaton*.

And since then I found more evidence in the commendation that I was right and he was wrong. And generally the argument of Freud, he never mentioned it, but, dual psychoanalytical patients or neurotics always speak, always bring out what they know – either they don't disclose it or they don't know it!

No wonder. On the same back Strasse, back street, where he lived, lived another man who was four years his younger, this was Theodor Herzl. Theodor Herzl, as you know, is known as, well, the originator of the political Zionism. But it is not exactly so. The originator was Moses Hess, who preceded him by 55 years, who was originator of Communism and of Zionism. Engels learned from him communism and brought Marx into the fold. And later Hess, who wrote *Rome and Jerusalem*, left, well, parted with communism and started Zionism. Hess, however, before he – I would say, when he – wrote his *Judenstaat*, "the state of Jews", he did not mention Palestine in the entire book. Only in answer to his book the Eastern Jews from Russia, from Poland, responded they found somebody who can speak in their name and they made him Zionist. And need to be said, before he came with the idea of Jewish state, Herzl had an idea to convert all Jews to Catholicism and with this purpose he wrote to the Pope.

Well, I thought I would send my interpretation of his dreams to Freud. Freud reached through Paris to London. Nobel Prize was denied him though he tried and let some other try, but the city of London gave him some honors, made him a citizen of the town, brought him a key to the town. But when I was about to send him this Dreams of Freud reinterpreted, the message came that he died. So I don't know would I have sent a few weeks earlier whether he would have been able to read it – I don't know.

Eight months I continued in New York, mostly in the public library on 42nd Street on the theme, Oedipus and Akhnaton, because these were the heroes of Freud and the book should be about him and about his heroes. About Moses I had not yet enough to talk.

But on the day we intended to travel back – eight months sabbatical, so to say, were over – Italy was not yet in war and we had our passage on an Italian ship, "Roma" (Planes were not flying yet across the ocean) – when I went to pick up my manuscript from a publisher, just close by here, the wife of the publisher said, "But we decided, we wish to publish this book. Don't take it away." But when I stayed and I returned after a number of days, they told me I have first to finish the book then they would decide. So I stayed already because a new idea very soon came to my mind in developing. The story is longer and I will not go into it.

Result was next 10 years, from 1940, spring, to 1950, spring, in library, mainly Columbia University library on 116th Street in various departments and after six or seven years, World War II was going on, students were put in uniform and drilled on the campus. I noticed that the one division of the library that I was not visiting was of psychology. Geology and archeology and astronomy and paleontology and Egyptology, all I was visiting. But not of psychology.

In the beginning, first year when in the first eight months between the summer of 1939 before the World War started, I was interested also in electroencephalograms because the idea to use it on epilepsy happened to be mine and it was printed as a preliminary communication in the paper that I told you, called *Physikalische Existenz der Gedankenwelt (Of the Physical Existence of the World of Thoughts)*, with the preface of Bleuler. So I was visiting in Boston medical places where they had an electroencephalogram. I made my first electroencephalogram on epileptic in Tel Aviv with a cardiogram; there was no electroencephalogram.

But now from 1940 these ideas that were developed first in *Ages in Chaos*, and then half a year later in *Worlds in Collision*, and few years later in *Earth in Upheaval*, this book on stones and bones, occupied my mind. So, we were four people, four students. She was starting high school, sneaked through though she was not prepared yet.

Ruth: You better explain that ...

My wife was a violinist who led the Palestine String Quartet for the several years we were in Palestine, but here she went as a student to study in Columbia University for the next five years sculpture and exhibited with remarkable success. And I was again student for those years.

However, since 1945, after the first five or six years, I started to see patients. Just could not make it in living, at that most necessities. The man who was sending me patients was Professor Dr. Paul Federn, that I mentioned before, who was previously president of Psychoanalytical Society in the days of Freud, and with whom we met that evening when we were only two on one side, on the side of Freud, against all his pupils in that meeting that I mentioned in the spring of '33. As he wrote to the president of the American Psychoanalytical Society who inquired about me for certain reason – literary reason – he answered him that he was sending me his most difficult patients and in his opinion I succeeded. I brought a copy of this letter, too.

Now soon I found out that the son of Federn was an Egyptologist, who studied in Vienna under Junker, and a few years later I told him about my ideas about ancient history, which I delineated in *Ages in Chaos*, of which the first volume was published in 1952, two years after *Worlds in Collision*. They had only initial common points. Later they diverged into two works.

Stekel also left Vienna. He was, like antenning himself, very sensitive. And in 1933 when I was there – it was the time when Hitler, only two months since he became chancellor, since he became the leader; before this he was under, what was it, the president, Hindenburg. Hindenburg was the president, he won over Hitler but then he gave Hitler all the power. Now Hitler since January of '33 was in power completely. And Stekel whom we were visiting at his home – my wife played with him a quartet – he was already then, five years in advance of the occupation of Vienna by Nazis, full of vibrations, so to say. He counted the time, the day. No wonder, because the Nazis were walking in Vienna with their flags and their signs and with music and what not.

Freud, before he left for England, well, also had to feel, finally, the situation which he would not like to recognize. Just didn't like to think about it. And even in his correspondence with Arnold Zweig, a German writer who lived at that time on Mount Carmel in Palestine, he expressed himself – I read it recently rereading this correspondence – whether Zweig thinks that still time is to find a refuge under the cross of Catholicism. So I was not entirely wrong in my analysis of these dreams.

Stekel, a generous man in every respect with his ideas, not a very careful writer. He could write so many books only because he was writing them and probably hardly read them before they were printed. He has his faults. One of his faults was that he could not hide his feelings. When I came to this country, I believe it was, we came in summer. Next summer – I go back now in time – my wife and myself after sending our children to a children camp, went to Lake George on the eastern side, just developed, and had a small cabin. And Professor Jelliffe, who was editor of *Psychoanalytical Review*, before this already told me that he and especially Dr. White, the other editor of the *Psychoanalytical Review*, who by that time was already dead, were great admirers of my work, of papers that they republished in the *Psychoanalytical Review*, reprinting them from German. The translator was at that time student in his last year, but later Professor of Psychiatry at Yale, Coleman. Dr. Coleman, three works of mine he translated. But the one about dreams of Freud was written – was translated, I would not dare at that time to write English myself, but it was translated by a lady whose original language was German, not English, and so today I would have done it better. And so told me Dr. Jelliffe, that summer on telephone, he regrets he cannot invite me to his estate there on the West Coast, West side of Lake George, because his wife knows that I was associated with Stekel, and she in Vienna went to Stekel and after she told him something intimate – I don't know what it was – he burst into loud laughter and she could not forgive him this and anything connected to Stekel is already taboo. So I stayed that summer in that little cottage, my wife was typing my *Ages in Chaos*. *Worlds in Collision* was not yet ready in my hand and, well and she figured that we can have it published by Passover time. She figured, and I was angry. So long to wait! Now, almost 40 years later – at least 38 years later, I have only published three volumes of this *Ages in Chaos* out of five. So, how mistaken can a man be?

I would not go into my experiences on my own, on my work because I wish to dedicate what we are talking today to this theme about those psychoanalysts whom I knew. So, Freud died, when he was 83. Stekel killed himself. And Federn, whom I described before, who supplied me with patients, he suicided too. Actually, I was at his home and spent the evening that night. My *Worlds in Collision* just appeared and two weeks after appearance was already on second place in New York Times, or Herald Tribune, and he was very happy for me. And he asked me to supply his son, who was kind of a recluse, who knew very much, but was a completely neurotic man. He knew Egyptology, bibliography of Egyptologists like nobody else, and actually prepared it during the war years for the Vatican to print it. Well, the next day I understood why he asked me to take care of his son, because he killed himself, having just lost his wife, and then having gone through his surgery of prostate that was cancerous and had to go for the second surgery. Probably he did not wish. There was collaboration with younger Federn till he died. He died actually of kind of neurosis that refused to eat most of human food. Well, such chance meeting played a great role because my *Ages in Chaos* would be by far not what it is, neither *Oedipus and Akhnaton*, if I would have not met Walter Federn, the son of Paul Federn. He died in '67, actually starved himself.

So the Second World War was to its end and for me the question, Why war? received a now different answer than I gave at the International Psychological Meeting Congress in 1937 in Paris. Actually, shortly before World War II – couple years before World War II – there was exchange of letters between Einstein and Freud. Einstein was younger than Freud. Einstein was born in 1879, in March, and Freud was born in, I believe, May 1856. So there was a difference of 22 years. In this exchange of letters, Einstein asked Freud, What could be helped?, What is the way to help? Freud could come with nothing as an answer. He didn't give even an answer. This is published as a booklet. It's well known this correspondence. These destructive urges in man will probably always exist – this was Freud.

Now, I must say, today Freud has a great following all around the world. Just I read, the other week in the *Science* magazine, an answer on kind of a questionary directed by the magazine to a number, maybe a dozen of prominent scholars, actually scientists, physicists mostly – and one of them said about Einstein: "I don't know who later would be regarded as the first, the greatest, Freud or Einstein." So, here is Freud, who in his time was debased – he was not even reviewed properly when his first book appeared – this dream book. I read also recently the letters that he was writing to Theodor Herzl, who was living on the same street, this I mentioned, but was writing for *Wiener Freie Presse*, I believe this was the magazine – daily – for which he later went to Paris to cover the Dreyfus affair and wrote his ..., at that time wrote his *State of Israel*. At this time Freud asked, almost begged Herzl to prepare a review of his book.

Now, the movement is great. It has certain reason now to have many followers because it became a profession. People earn their living where this kind of philosophy is certainly secured from, well fate of philosophy, where no such dream is supplied. On the other hand you know that many followers of Freud, his pupils, dissipated their thoughts, prepared their own ideas and disagreed with Freud, practically on every point. So Jones writing his biography had to find who still remained faithful to him. So people like Eitingon who had a clinic in ... helped to open a clinic, analytical clinic in Berlin and later came to Palestine with the rest of these German psychoanalysts, visited me, too. But not many. However, those who follow Freudian theory and practice, do not follow him all the way. Actually they stop in the middle. And this group, not different from the Analytical Society of America, just the same.

I had, in November, a group, like your group. They are mostly, all of them almost medical doctors with ... one of them was Professor of Psychiatry from Denver, they coming together every year in Princeton and they wished to hear me. So I lectured to them. This is only time I lectured – my lecture before this was '75. So you have – at 80 I stopped to lecture. However there would be an exception – in a couple weeks in Princeton I will speak on Sciences and Humanities of Tomorrow.

Now, you know, of course, that Freud believed that between the age of 2 and 5, this age most neuroses are, so to say, planted in the human soul. Later they may have a long period of no open progress, and then some time later burst through. But you have to observe that he changed his views, maybe partly in competition with Jung. And this need to be followed through better than I did it. Because in my book that I am preparing now, writing now, *Mankind in Amnesia*, I dedicate a chapter to Freud and to Jung and I already received a letter from one Jungian, long letter, claiming, debating with me the problem, trying to show that Jung was the first. I find that Freud had it already in 1913, *Totem and Taboo,* the idea. And what the idea is? The idea is that, actually the trauma, traumatic experience that causes the oblivion, the forgetfulness, the amnesia, whether it is psychological trauma or physical trauma, it – Freud claims – it is not only with the individuum, the person whom you have before you on the couch, it goes philogenically to the beginning of human race. He developed this idea again and again and, I must say, even Jones, his biographer, almost accepted only that it is from his three volume book. But if you take his last book of Freud, *Moses and Monotheism*, the second and third chapters – this makes the book where he doesn't speak already so much about Moses, but about other things – this is actually repeated again and again, this claim. He also offered Ferenczi, one of his close collaborators, to write together a book knowing well that science – biological and neurological and physiological science of his days, like also of days of today almost – would object inheritance of acquired characteristics. So, if something happened to the person, a trauma, it would be not transplanted through the next generation and the next generation. But he wished to challenge this. And his book – I took a few pages from my *Mankind in Amnesia* to have before me some words from Freud on this from his various books – in one place he writes that way:

"We must finally make up our minds to adopt the hypothesis that the cyclical precipitates of the primeval period became inherited property which in each fresh generation called not for acquisition, but only for awakening. We find that in a number of important relations children react not in a manner corresponding to their own experience, but instinctively, like the animals in a manner that is only explicable as philogenic acquisition."

And again in other place: "The reader is now invited to take the step of supposing that something happened in the life of the human species, similar to what occurs in the life of individuals, of supposing that is, that here too, events occurred of a sexually aggressive nature, which left behind them permanent consequences, but were, for the most part, fended off and forgotten and which after a long latency came into effect and created phenomena similar by symptoms in their structure and purpose."

I will maybe make two quotes more:

Psychoanalytic Lecture

"Some portion of the cultural acquisitions have undoubtedly left a precipitate behind them in the Id – I-D – Id, much of what is contributed by the super-ego will awaken as an echo in the Id, not a few of the child's new experiences will be intensified because they are repositions of some primeval philogenetic experience."

Ruth: Aba, we have about 10 minutes.

Now the last I will read: "If we consider mankind as a whole and substitute it for a single human individual, we discover that it too, has developed delusions which are inaccessible to logical criticism and which contradict reality." I will stop here.

In my work that I am writing now, *Mankind in Amnesia*, I agree with Freud, but show also that Freud stopped before the end. He recognizes there were some traumatic experience to human race, but he did not know what the traumatic experience was. It was not the father of the herd in the Stone Age, killed by his grown-up sons who wished also to posses their mothers. There was a traumatic experience, a traumatic experience on a global scale and in historical, also prehistorical times. On some of them I wrote in books like *Worlds in Collision* and *Earth in Upheaval*. And without knowing our past, being even afraid to look into our past, the man is a victim of amnesia – the human race, victim of amnesia, playing with thermal nuclear weapons. And this is maybe the main and the most important part of the entire analytical teaching. So saw it Freud, even not recognizing what was the trauma. So, thank you.

Question: I would like to, if you could give me your opinion of why in the whole field of psychoanalysis, economic and political conditions are completely absent in dealing with any individual. It is like the individual existing in vacuum and none of the conditions surrounding seem to be paid attention to?

Answer: I would not say they are omitted, but they are more of them represented as substitute for sexual phenomenon or sexual activity. So this is a point that, for example, Adler would make a strong argument of it.
Anybody else have any question?

Question: You mentioned that you met Freud – you met Freud at his 77th birthday and you had about 30 minutes of interview with him. Did you meet him any other time?

Answer: No, corresponded only, we corresponded and I mentioned also he published several papers of mine in his magazines. But of correspondence it goes to 1921 or '22 to his death.

Question: Have you published this correspondence?

Answer: I gave most of the letters to Freud's archive but here I brought one of the letters in his handwriting – his letters are always by hand.

Question: What was the association of the environment in the type of patients that you experienced from Europe opposed to the United States?
Are patients different in Europe than here?

Answer: Well, in Palestine I had many. Here I had few. As you have heard some difficult ones. Everyone is different. I have never seen two patients the same.

Question: Was there a cultural difference?

Answer: Well, I have not selected my patients by their culture and there were people very simple, very primitive, and people very sophisticated. Here and there.

Question: I have one question. You had said, Dr. Velikovsky, that Freud had indicated that possibly he might have been better off toward the end by having converted to Catholicism. You said that Freud indicated that he – to his benefit to have converted to Catholicism. I know that in your article you were talking about his dreams as being – he was concerned about the crucifix, he was concerned about other affects upon his life. Had he held these ideas for long? Had he discussed this with you privately?
Had Freud professed to you, personally, that he wished to convert to Catholicism?

Answer: No, never. Not in writing, and not the only time that we were together.

Question: When you were together he talked about that with you?

Answer: No, maybe we spoke about Jewish problems because I came from Palestine and he wrote before this that he would like to come to Palestine, more than to any other place.

Question: Who was the person who most influenced you in your life?

Ruth: I'd like to know the answer, too.

Answer: I would say ...

Ruth: You have to make it about two minutes.

I'll answer in two minutes. Most probably most influence came from my father. But, let me say so: I was born on 10th of June, 1895. At this time three events took place.

Physics entered a new stage, completely new stage with the discovery of Roentgen 1895, Becquerel 1896, then Curie in 1898, and then already Townsend ..., quantity and quality, Plank and so on. This was the beginning, in one way. Politically as I mentioned before, on the 10th of June – reading the diary of Theodore Herzl written in a hotel near Magdalene in Paris – on that day that he hears a noise of wings over his head, as if an angel guides him in writing that book – this his feeling was when he started to write. Even Palestine was not mentioned. And, you know, last week the president of America came to put flowers on his grave in Jerusalem. At that time it was not expected. Now, and third, Freud wrote ... Freud started and wrote with Breuer his first paper on psychoanalysis. So, these three ... occupation by these three people, in some way, if you wish, predestined my own occupation in the future.

And in the school, I believe the teacher of Russian composition, impressed me most and guided me most. And he believed I will be a Russian poet. I would have disappointed him.

Well, and then I was given the good fortune of having most dedicated wife and daughters. I don't think that everyone is blessed with this blessing, that helped me through 29 years now of opposition by scientific community. Now it's – you feel it with Venus and Jupiter probes that went all my way and all against their way, that there is a kind of uneasiness, at least I feel it by letters and by invitation from press, from television stations, from universities and from others. But, through these 29 years of opposition preceded by 10 years of study in the libraries I was supported and could carry through, not without some, well, mental wounds myself, I would say, traumatic experiences were there.

Moravian College 11/11/71 (Gary A. Becker)

Aba – Ima's sculpture

"Velikovsky" in Hebrew

please wait
shall return
soon.   I.V.

Around the Subject

The Author

Dr. Ruth Velikovsky Sharon learned at the desk of her distinguished father, Dr. Immanuel Velikovsky, a prominent psychiatrist and eminent man of science whose genius engaged even the mind of his friend and contemporary, Albert Einstein.

Dr. Sharon received a B.A and M.A. degrees from New York University and a Ph.D. from the Union Institute and University. She was a graduate of the Center for Modern Psychoanalytic Studies and a certified psychoanalyst.

She died in 2012.

Books by Ruth Velikovsky Sharon:

- *Aba – The Glory and the Torment* (1995)
- *I Refuse to Raise a Brat* (Co-author with Marilu Henner, 2000)
- *The Truth Behind the Torment* (2003)
- *Shame on You – You Were in My Dream* (2003)
- *The More You Explain ... The Less They Understand* (Co-Author with John C. Seed, M. D. 2005)
- *Imagine Art* (2009)
- *Insights of a Psychoanalyst* (2011)

Immanuel Velikovsky
–
The Truth Behind the Torment

Ruth Velikovsky Sharon, Ph.D.

ISBN 978-1-906833-21-3
(Softcover)
978-1-906833-61-9
(Hardcover)

In this supplement to her father's biography, Ruth Velikovsky Sharon, Ph.D. depicts the true facts about the campaign against him.

She publishes revealing letters in full length, that show the true nature of the undeserving – unscientific – treatment of Velikovsky by the scientific establishment, a treatment that appears rather medieval than enlightened.

Also included is a chapter written by plasma physicist C. J. Ransom, Ph.D. dealing with Velikovsky's theories and the controversy about our Solar System, the ancient sky and the role of electromagnetism in cosmology.

Shame on You – You Were in My Dream

Ruth Velikovsky Sharon, Ph.D.

ISBN 978-1-906833-01-5

Finally a new and easy guide to the understanding of dreams, which really makes sense!

Ruth Velikovsky Sharon, PhD has developed a completely new understanding of the nature of dreams, which is fascinating because of its simplicity and its practical orientation.

She questions ideas we have long taken for granted. She asks us to reconsider what the word "dream" really means. She shows us that to use the word "dream" in partnership with "He is a dreamboat" or "My dream house!" is to misuse or even abuse the word "dream".

In her book, Dr. Sharon describes the way that parents can be of help vis a vis dreams: Listen and Learn. Ask your children how they felt in the dream, ask them what they thought in the dream. She includes chapters on manipulation in dreams, dream catchers and other gadgets and the environment and dreams.

Also included is a reprint of the article »A New Understanding of Dreams«, published by Dr. Sharon in *New Jersey Medicine*, Journal of the Medical Society of New Jersey, January 1995 issue.

The More You Explain ... The Less They Understand

Ruth Velikovsky Sharon, Ph.D.
John Cathro Seed, M.D.

ISBN 978-1-906833-00-8

In this, perhaps the most encompassing of her works, Dr. Ruth Velikovsky Sharon brilliantly lifts the veil that shrouds the mystery of psychoanalysis, revealing intrinsic truths that can forever assist us in our journey to self-discovery and growth.

Like a finely tuned and well-trained instrument, Dr. Sharon makes her probe into the human psyche sound easy – resulting in a compilation of luminous insights that are warm in their humanity, vibrant in their simplicity, and even touched with humor.

Harvard Medical School Graduate, Dr. John C. Seed's contribution of the Physical Health chapter will enlighten the medical community as well as the average reader, and if abided by, will help prolong life.

 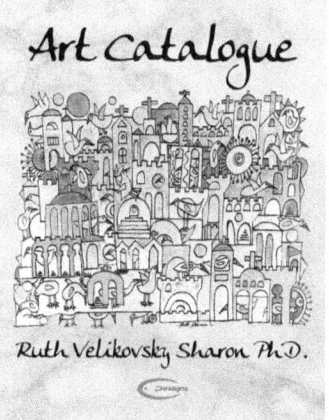

Imagine Art   **Art Catalogue**

Works of Art by
Ruth Velikovsky Sharon, Ph.D.
and Elisheva Velikovsky

Ruth Velikovsky Sharon, Ph.D.

ISBN 978-1-906833-02-2   ISBN 978-1-906833-03-9

The name of Velikovsky is mainly known from the scientific and historical discoveries of Immanuel Velikovsky.

Far less known is the artistic dimension in the Velikovsky family, mainly expressed by Elisheva (or "Elis") Velikovsky and Ruth Velikovsky Sharon, PhD., the wife and daughter of Immanuel Velikovsky. For everyone interested in and fond of visual and plastic arts these booklets will give an exhaustive overview of the remarkable range of the works of these two artists.

Worlds in Collision

Immanuel Velikovsky

ISBN 978-1-906833-11-4
978-1-906833-51-0

With this book Immanuel Velikovsky first presented the revolutionary results of his 10-year-long interdisciplinary research to the public – and caused an uproar that is still going on today.

Worlds in Collision – written in a brilliant, easily understandable and entertaining style and full to the brim with precise information – can be considered one of the most important and most challenging books in the history of science. Not without reason was this book found open on Einstein's desk after his death.

For all those who have ever wondered about the evolution of the earth, the history of mankind, traditions, religions, mythology or just the world as it is today, *Worlds in Collision* is an absolute MUST-READ!

Earth in Upheaval

Immanuel Velikovsky

ISBN 978-1-906833-12-1
978-1-906833-52-7

After the publication of *Worlds in Collision* Immanuel Velikovsky was confronted with the argument that in the shape of the earth and in the flora and fauna there are no traces of the natural catastrophes he had described.

Therefore a few years later he published *Earth in Upheaval* which not only supports the historical documents by very impressive geological and paleontological material, but even arrives at the same conclusions just based on the testimony of stones and bones.

Earth in Upheaval – a very exactly investigated and easily understandable book – contains material that completely revolutionizes our view of the history of the earth.

For all those who have ever wondered about the evolution of the earth, the formation of mountains and oceans, the origin of coal or fossils, the question of the ice ages and the history of animal and plant species, *Earth in Upheaval* is a MUST-READ!

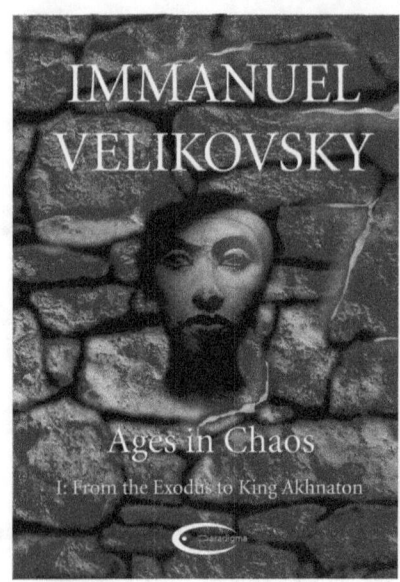 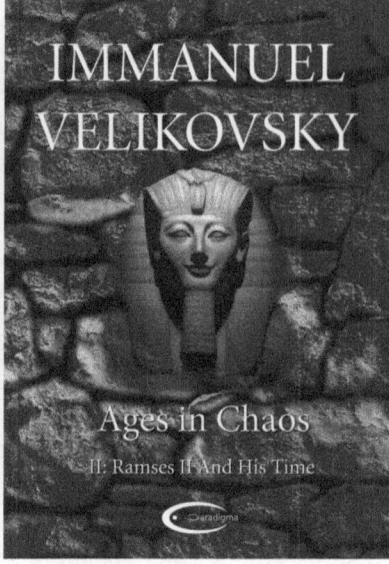 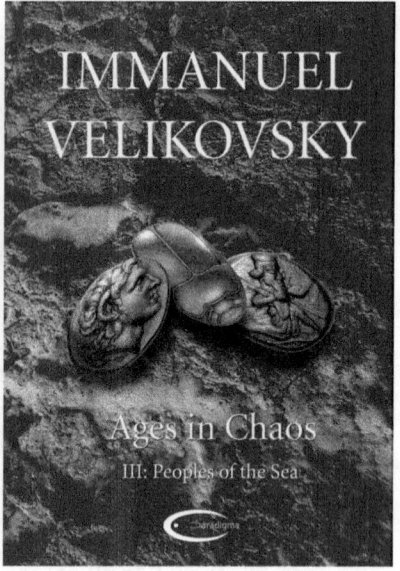

Ages in Chaos

Immanuel Velikovsky

I: From the Exodus to King Akhnaton
ISBN 978-1-906833-13-8
978-1-906833-53-4

II: Ramses II and His Time
ISBN 978-1-906833-14-5
978-1-906833-54-1

III: Peoples of the Sea
ISBN 978-1-906833-15-2
978-1-906833-55-8

In his series *Ages in Chaos*, Immanuel Velikovsky undertakes a reconstruction of the history of antiquity.

With utmost precision and the exciting style of presentation typical for him he shows beyond any doubt what nobody would consider possible: in the conventional history of Egypt – and therefore also of many neighboring cultures – a span of more than 600 years is described which has never happened! This assertion is as unbelievable and outrageous as the assertions in *Worlds in Collision* or *Earth in Upheaval*. But Velikovsky takes us on a detailed and highly interesting journey through the – corrected – history and makes us witness, how many question marks disappear, doubts vanish and corresponding facts from the entire Near East furnish a picture of overall conformity and correctness. In the end you not only wonder how conventional historiography has come into existence, but why it is still taught and published.

In an extensive supplement to *Peoples of the Sea* Velikovsky delves into the fundamental question of how such a dramatic shift in chronology could have come about.

In a further supplement he discusses the very interesting conclusions that can be drawn from radiocarbon testing on Egyptian archeological finds.

Just as Velikovsky became the father of "neo-catastrophism" by *Worlds in Collision*, he became the father of "new chronology" by *Ages in Chaos*.

Mankind in Amnesia

Immanuel Velikovsky

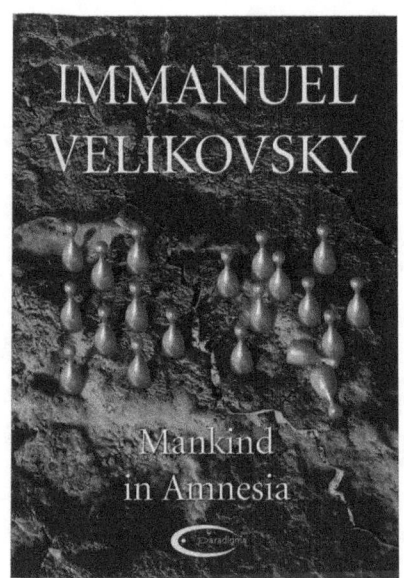

ISBN 978-1-906833-16-9
(Softcover)
978-1-906833-56-5
(Hardcover)

Immanuel Velikovsky called this book the "fulfillment of his oath of Hippocrates – to serve humanity." In this book he returns to his roots as a psychologist and psychoanalytical therapist, yet not with a single person as his patient but with humanity as a whole.

After an extremely revealing overview of the foundations of the various psychoanalytical systems he makes the step into crowd psychology and reopens the case of *Worlds in Collision* from a totally different point of view: as a psychoanalytical case study. In this way he shows that the blatant reactions to his theories (which are still going on today) have not been surprising but actually inevitable from a psychological perspective – which equally holds for those who have defined our view of the world. At the same time he is able to reclassify the theories of Siegmund Freud and of C. G. Jung finding a common basis for them.

A journey through history, religion, mythology and art shows the overall range of the collective trauma and gives us – the patients – a message of extraordinary urgency and importance for the future.

www.ingramcontent.com/pod-product-compliance
Lightning Source LLC
Chambersburg PA
CBHW080423230426
43662CB00015B/2197